12-13-21 ERH 4.95

D0049616

True
or Poo?

True or Poo?

The Definitive Field Guide to
Filthy Animal Facts and Falsehoods

Nick Caruso and
Dani Rabaiotti
Illustrated by Ethan Kocak

NEW YORK   BOSTON

Hachette Books
Hachette Book Group
1290 Avenue of the Americas, New York, NY 10104
hachettebooks.com
twitter.com/hachettebooks

Originally published in 2018 by Quercus in Great Britain.

First Edition: October 2018

Hachette Books is a division of Hachette Book Group. Inc. The Hachette Books name and logo are trademarks of Hachette Book Group, Inc.

The publisher is not responsible for websites (or their content) that are not owned by the publisher.

The Hachette Speakers Bureau provides a wide range of authors for speaking events. To find out more, go to www.hachettespeakersbureau.com or call (866) 376-6591.

Library of Congress Control Number: 2018952707

ISBNs: 978-0-316-52812-2 (hardcover), 978-0-316-52809-2 (ebook)

Printed in the United States of America

LSC-C

10 9 8 7 6 5 4 3 2 1

For Science Twitter, a community whose enthusiasm, support, and awesome animal-related content made this book happen. Thanks for letting people know science is for everyone.

INTRODUCTION

Myths about animals are as old as time. For example, salamanders were once believed to be born from fire or even immune to it, because people observed salamanders emerging from logs that they were burning to keep warm. In reality, these salamanders make their homes inside logs and are very sensitive to rising temperatures. When their homes begin to heat up, the salamanders quickly evacuate. Another common myth is that ostriches bury their heads in the sand to "hide" from predators. (Both of these myths as well as others throughout this book can be traced back to the naturalist and philosopher, Pliny the Elder.) In reality, however, observations of ostriches "burying" their heads was likely them lowering their head into their nests, which consist of holes in the ground, to turn their eggs, or potentially just moving their head to the ground to feed.

These myths, however, can't compete with some of the weird and wild things that animals actually do every day. Animals and their evolutionary adaptations are amazing, we don't need made-up facts about them. For example, did you know that salamanders can regenerate their tails as well as their arms and legs? And ostriches can run at speed of over 70 kmph? While it may seem innocuous to believe these animal myths, often it can lead to an underappreciation of how truly great the natural world is or, worse, needless harm toward certain species that have obtained a bad reputation as a result of this "fake gnus."

As zoologists, we have had the privilege to get up close and personal with a wide variety of animal species, as well as reading thousands (and we seriously mean thousands) of papers on animal-related topics throughout our careers. On top of this, we hang out with a lot of other people who study and work closely with various animal species, who have many weird and wild animal tales to tell. It turns out, in fact, that nature is pretty gross—the Victorians may have had some very ingrained ideas about the beauty of nature, but in reality animals do a lot of really, really disgusting stuff.

In this book you will find a mixture of bizarre myths that many people believe about the animal kingdom, along with some absolutely unbelievable facts. They cover topics from mating and parenting to feeding and digestion and strange names, and the peculiar places animals call home. All with a gruesome and gross slant, of course. Can you guess whether they are True… or just a bunch of Poo?

CONTENTS

ANIMAL EATING HABITS 24

DIGESTION AND EXCRETION 50

DEFENSE

COURTSHIP, MATING, AND PARENTING

In order to persist, species need to reproduce. However, reproduction in the animal kingdom is incredibly variable—one animal's stink might be another's kink! There is so much that needs to happen in order for reproduction to be successful. First, animals need to find and attract a mate; this is the courtship stage. Typically, males compete to win over females and this can involve elaborate displays with a healthy dose of swagger which can often relate information about the male's reproductive fitness to the females. For example, a bright, colorful, and long tail in peacocks signals to the female, known as a peahen, that the male is in good health and quality, because he can acquire abundant resources to devote to his plumage and still escape predation while being very clearly visible. But sometimes a physical encounter between males is needed for the female to decide who should sire her offspring. As always in the animal kingdom, though, there are exceptions, and in some animals, such as the *Jacana*—a genus of large-footed wading bird—the females are the ones that fight to secure the best mate around.

During mating, gametes, which are the reproductive cells that contain half of the DNA of each parent, unite to form a zygote. In humans, development of the zygote, known as gestation, on average lasts 280 days and occurs within the female, but this is by no means a constant throughout the animal kingdom. After an offspring is born or hatched, some species are content to let their progeny fend for themselves, but it might surprise you to know that there are also some attentive and dedicated animal parents. Some animals work together to raise their young, but sometimes one parent does all the heavy lifting—often this is the female, although males do occasionally lend a helping hand, fin, or claw.

In this section you will find the answers to questions about all the weird and wonderful ways by which animals reproduce—from courtship displays through mating and finally to parenting. What do hooded seal courtship and children's birthday parties have in common? (See page 5.) Do some species really find urine attractive? (See page 7.) Which animals use their genitalia in a way that resembles Olympic sports? (See pages 8 and 10.) Are birds really as romantic as they seem? (See page 15.) Or will they abandon their young if something is amiss? (See page 17.) What really happens to male clownfish when their female partner meets a terrible fate? (See page 18.) What is so special about the Surinam toad? (See page 19.) What do infant—mammals get up to when they are born—apart from being adorable? (See page 6.)

The next few pages will also highlight just how far animal parents will go to raise their young. Are females the only ones that practice parental care? (See page 13.) Just how big a ball of poo can a dung beetle push? (See page 11.) Will they eat their offspring's waste (see page 20) or let their offspring eat them (see pages 14 and 22)?

HOODED SEALS USE BALLOONS TO ATTRACT MATES

TRUE OR POO? TRUE

When humans think of balloons, they might think of birthday parties, but when a female hooded seal (*Cystophora cristata*) sees a balloon, she is more likely to think of mating. Hooded seals are so-named because of the presence of a "hood"—an inflatable bladder between the eyes and upper lip, found on males. Males inflate their "balloons" to make themselves appear larger and more threatening toward other males, in the hope of exerting their dominance. But that's not all, male hooded seals also have an inflatable nasal membrane that appears bright red or pink when inflated. If the display fails to intimidate, males may resort to fighting to decide on their social standing. But his work isn't done yet—he also presents his balloon to a female, hoping to impress her enough that she will agree to mate with him. Given this rather strange mating ritual, we do not recommend that you invite hooded seals to a child's birthday party.

INFANT HOOFED MAMMALS CAN WALK WITHIN MINUTES OF BEING BORN

TRUE OR POO? TRUE

Humans generally start walking a year after they are born, but giraffes, elephants, and other hoofed mammals accomplish this "feet" within minutes. This doesn't mean that humans are worse parents than other mammals, but rather it has to do with our brains. Humans have large brains for our size and subsequently a large cranium, or skull. For our species' existence this is a very good thing; we can use tools, solve complex problems, communicate, and even write books about farts and poo. But those large skulls need to be able to fit through the birth canal, so our brains are relatively underdeveloped at birth, which means that a good portion of our development happens afterwards. The brains of hoofed mammals, on the other hand, are relatively more advanced than humans' by the time they're born. Their gestation—the time between conception and birth—is typically longer, at around 14 months for a giraffe and as long as 22 months for an elephant. In fact, scientists have found that the time it takes for a given species of mammal to walk from conception (including gestation) is well predicted by the average size of that species' brain.

CAPUCHIN MONKEYS USE THE SCENT OF URINE FOR COURTSHIP

TRUE OR POO? TRUE

There are numerous similarities between humans and other primates, including our genetic makeup, our use of tools and our propensity to socialize with other members of our species. Likewise, we can find similarities between humans and capuchin monkeys (subfamily *Cebinae*), which can be found in forested areas in Central and South America. To get a male's attention, female capuchins will often grab or pull his tail or may even throw rocks at him; behaviors that are reminiscent of schoolyard children. However, one behavior that capuchins and (most) humans do not share is that male capuchins will urinate into their hands and then rub it into their fur. It may seem gross, but there's a good reason for this urine wash. Studies have found that females are able to detect testosterone levels in the urine, which allows them to identify sexually mature males and differentiate potential mates based on their social status. At this point, we are unaware of a similar ability in humans to use the scent of urine to distinguish potential mates, so we advise against using a urine-based aftershave.

FLATWORMS FIGHT AND STAB EACH OTHER WITH THEIR GENITALIA

TRUE OR POO? TRUE

The interactions between species can often be described as an "evolutionary arms race," in which evolutionary adaptations in one species are eventually countered by adaptations in another. For example, rough-skinned newts (*Taricha granulosa*) have evolved the secretion of a toxin that can cause paralysis and death, but some garter snakes (*Thamnophis sirtalis*), which eat newts, are resistant to it. But for many species, like flatworms (phylum *Platyhelminthes*), an ongoing evolutionary arms race is happening *within* a species.

Hermaphroditic flatworms, where each individual is equipped with both egg- and sperm-producing organs, will fence each other with their penises, which are extendable and sharp—the longer and sharper the better. The winner will then stab the loser and inject them with their sperm. This is known as traumatic insemination, and it can be observed throughout the animal kingdom in species such as rotifers, gastropod snails, nematodes, and the giant squid. Traumatic insemination has evolved in species that had pretty gross mating behavior in the first place, like males that glue the female's reproductive tract closed, or block it with their own reproductive organs in the hope of preventing other males reproducing with her. By piercing and injecting the female with their hypodermic genitalia, competing males can then sidestep this barrier.

9

OCTOPUSES HAVE DETACHABLE PENISES

TRUE OR POO? TRUE

What would you do if the female of your species was five times your size and could strangle and eat you when you attempted to mate with her? If you answered "detach your penis and throw it at her," you might just be an argonaut (genus *Argonauta*). Argonauts are a genus of octopuses that can be found in the open ocean, and they are distinguished from other octopuses in that they produce their own shell. The male argonaut's "penis" is actually a modified arm, known as a hectocotylus, that contains sperm. When removed from the male's body, this appendage continues to swim toward the female, attaches to her mantle (the structure behind their head where you can find their organs), and can be stored in her mantle cavity. Unfortunately for the male, he dies shortly afterwards (so would you, if you ripped one of your arms off and floated off into the ocean), so he only gets one shot at siring offspring. Females, on the other hand, can store sperm and fertilize her eggs from multiple males.

While it might seem to be a rather unfortunate fate for the male argonaut, the alternative is being eaten without the chance to pass along his genes to the next generation, which is even less ideal.

DUNG BEETLES CAN PUSH A BALL OF POO OVER 1000 TIMES THEIR OWN BODY WEIGHT

TRUE OR POO? TRUE

This awesome superfamily of insects—the scarab beetles or dung beetles, *Scarabaeoidea*—all feed partly or exclusively on feces. It may not sound very appetizing, but they do an incredibly important job of recycling nutrients—without them there would be poo all over the place! And that would lead to an increased spread of disease and poor soil quality. There are three types of dung beetle: dwellers, who just hang out in the dung where they find it; burrowers, who tunnel under the dung and bury it; and rollers, which form the dung into balls before rolling it away to their burrows and breeding chambers. This third group go to some extreme lengths to get this poopy food ready for their babies to eat. The male shapes the feces into a huge ball and starts rolling it toward soft ground. If the female likes what she sees (and who wouldn't?), she will climb onto and ride the ball—adding her own weight to what he is pushing (helpful!). Once they get to a good spot, they both bury the ball and prepare it for brooding, then the female lays her eggs in it. One species in particular, *Onthophagus taurus*, is particularly good at pushing around big balls of poo—males can push up to 1142 times their own weight. That would be the equivalent of a human rolling a ball the weight of a sperm whale. Talk about a strong father figure!

MALE SEAHORSES GET PREGNANT

TRUE OR POO? MOSTLY TRUE

For all 54 known species of seahorses (genus *Hippocampus*), the male is observed giving birth to the young, but does that mean the male gets pregnant? While we have rated this statement as "true," male seahorses don't become pregnant in the manner in which most people are familiar—they aren't actually producing eggs. Male seahorses are equipped with a brood pouch on their abdomen, and after successfully courting the female with a dance, she will deposit eggs into this pouch. While scientists don't know the exact cause for this parental role reversal, it is possible that this trait evolved because it allowed females to produce more eggs; handing over the reins to the male means more time for her to devote to producing the next batch. After being fertilized by the male, the eggs will develop within this pouch and eventually he will give birth to the young – which are fully formed, tiny versions of the adults.

Male seahorses, however, contribute more than just a pouch for developing embryos, studies have shown that they provide nutrients, remove waste and even protect their young from infection. Interestingly, the genes that are expressed during a male seahorse's pregnancy are similar to those found in pregnant mammalian females. However, unlike mammals, seahorse young can develop outside of the brood pouch, although this process typically takes longer, and fewer offspring survive.

SOME ANIMAL OFFSPRING EAT THEIR WAY OUT OF THEIR MOTHER

TRUE OR POO? TRUE

The wonderful world of invertebrates, where for a number of species it is perfectly normal to, instead of giving birth, have your children eat their way out of your body. If you thought the back-bursting frogs (see page 19) sounded like a pretty rubbish way to reproduce, imagine being devoured from the inside out by your voracious children. Well, that is what lies in store for you if you are a female *Adactylidium*—a type of microscopic mite. But it gets weirder... you are born pregnant! When the mite's eggs (five to eight of them) hatch inside the mother's body cavity, the single male offspring mates with his female siblings. All this while they are still inside their mother! The young female mites then eat their way out of their mother's body to enjoy just four days of freedom before their own pregnant daughters devour them, and so the whole incestuous matricidal cycle continues. While four days seems short, the male is lucky to live even a few hours outside their mother's body, as his job is done and he has no further reason to live. Eating your own mother is known as matriphagy and is practiced by a wide variety of species, including some earwigs (see page 136), strepsipterans (a type of flying, parasitic insect) and caecilians (see page 22).

ALL BIRDS ARE MONOGAMOUS

TRUE OR POO? POO

Many people have observed male and female birds in pairs (such as ducks swimming together), often raising offspring together, which has likely led to the spread of the myth that birds are monogamous—that is, they only have one mate. Monogamy is a relatively rare practice in the animal kingdom, but for birds the number of monogamous species is high. Still, it's not the case for all birds: approximately 92 percent of the nearly 10,000 species of bird exhibit a form of monogamous or pair-bonded parents. Some species (such as albatrosses; family *Diomedeidae*) are monogamous for life, others for consecutive breeding seasons (such as mourning doves; *Zenaida macroura*), whereas other species may only be monogamous for a single breeding season (such as a bull-headed shrike; *Lanius bucephalus*) or just a single nesting (such as a house wren; *Troglodytes aedon*). Songbirds, in particular, are renowned for their pair-bonding. DNA analysis, however, has shown that the offspring's genes do not always match those of the male, or sometimes even female, that raised it. In blue tits, for example, up to 50 percent of nests contain the eggs of a male other than the one raising them: so really, birds aren't as faithful and romantic as they might seem...

IF YOU TOUCH A BABY BIRD
THEIR PARENTS WILL ABANDON THEM

TRUE OR POO? POO

Nope, no matter what you were told, touching a baby bird will not make its parents abandon it. Birds are rather fond of their offspring, and tend not to let them die if they can help it. The myth that birds will abandon their chicks if you touch them comes from the belief that birds can smell the scent of humans on their offspring. In reality, however, birds have a pretty rubbish sense of smell, and are unlikely to be able to tell if you have touched their young! Still, if you find a baby bird, you shouldn't touch it—this can be distressing to both the chick and its parents. The best course of action is just to leave it where it is and keep an eye out for the parents returning. If the bird is in immediate danger, such as stranded on a busy road or path, it is generally okay to move it a short distance out of harm's way. Disturbing a nest can be detrimental to the birds and yourself; soon after the eggs are laid parents may abandon them as they are able to lay another clutch somewhere that isn't disturbed by predators. On top of that, many species have a tendency to dive-bomb nest predators, which can lead to some nasty head wounds (the author definitely does not speak from experience...), so don't go poking about in birds' nests!

CLOWNFISH STAY ONE SEX FOR THEIR WHOLE LIVES

TRUE OR POO? POO

There are nearly 30 species of clownfish and, as you probably know, they all have one thing in common: a symbiotic relationship with anemones. The clownfish lives in the anemone and keeps it clean, and the anemone keeps the clownfish safe from predators with its stinging tentacles. Clownfish avoid getting stung by covering themselves in a thick mucus.

While getting pregnant might seem like a big step to maximize your reproductive output, clownfish take this one step further: they change sex. All clownfish live in groups with a strict dominance hierarchy, where the largest male and the largest female mate and produce eggs. The female, as with most fish, is typically larger than the male, as she has to carry up to 1000 eggs at any one time. If the female dies, her male partner will become female and the next largest male will take his place.

This behavior, known as sequential hermaphroditism, is by no means unique to clownfish; moray eels, wrasse, and gobies have all been observed changing sex to increase the number of offspring produced throughout their lives. They are far from the organism with the most complicated sexes though, as some species of fungi have been found to have over 36,000 of them!

SURINAM TOADS GIVE BIRTH THROUGH THEIR BACKS

TRUE OR POO? TRUE

The common Surinam toad, *Pipa pipa*, is a species of aquatic frog found in South and Central America. They are probably best known for their rather unusual, and gross, method of giving birth. During the breeding season the male will make a loud clicking sound to attract a mate. Once a willing female arrives, they flip about and somersault through the water together, with the female laying up to 100 eggs and the male fertilizing them. During this process the eggs become stuck to the specially thickened skin on the female's back. These eggs gradually become more and more embedded in her skin over the following days, in a disgusting but innovative method of keeping them safe from predators. As with most frogs and toads, the eggs hatch into tadpoles; however, this happens inside the mother's back! Once the offspring reach the froglet stage (small versions of adult frogs) they break out through the mother's skin and pop into the water—kind of like popping a big pimple, but instead of pus a baby frog comes out. Ah, the beauty of motherhood. If that wasn't weird enough for you, Surinam toads also have no tongues.

BIRDS KEEP THEIR NESTS CLEAN BY EATING THEIR CHICKS' POO

TRUE OR POO? TRUE

Imagine a bird's nest. Is it covered in poo? Probably not. If it is it might be a booby (page 109) or you maybe need to get out and look at more bird nests (from a safe distance of course—don't disturb the nest), because they are generally impressively clean. So how is it that birds keep their nests so tidy and poo-free despite the fact their offspring live in them? Unfortunately for human parents looking for some housekeeping tips, you may not like the answer. Many species of baby birds poo into a special thick mucus membrane called a fecal sac—kind of like a diaper for birds. This keeps the poo contained in a ball and stops it from soaking the nest. Baby birds will poo out these sacs within seconds of being fed, allowing the parents to clean them up. When the offspring are young the parents then eat these sacs, poo and all. It was originally thought that this was because the parents gained nutrition from the poo, however more recent research has suggested that the parent birds may just be lazy. As a bird parent with so many loud screaming mouths to feed, are you going to spend time and energy carrying these sacs of poo? Or are you just going to eat it? Turns out the answer depends on how big the fecal sac is—smaller sacs take up less room in the parents' stomach, so get eaten. As the young birds get older and the fecal sacs bigger, the parents stop eating their offsprings' poo and carry the fecal sacs away from the nest instead.

CAECILIANS EAT THEIR MOTHERS

TRUE OR POO? MOSTLY POO

Caecilians (order *Gymnophiona*) is a collective name for the 207 known species of legless, wormlike, and primarily burrowing amphibians that can be found in wet tropical regions of Central and South America, Asia, India and Africa. A caecilian adult's diet primarily consists of earthworms and arthropods, as well as other invertebrates. The diet of caecilian young, however, is quite different from that of the adults—they eat their mother's flesh (but not all of it). Some caecilian species (like the East African caecilian, *Boulengerula taitanus*) lay eggs, and those newly hatched individuals are born with specialized teeth that allow them to peel and tear away the outermost layer of skin from their mother, which is thickened and enriched with lipids. While caecilian mothers may lose up to 14 percent of their body weight to feed their young, their skin does grow back within three days. However, other caecilian species (like Seshachari's caecilian, *Gegeneophis seshachari*) give birth to live young, which means the skin-feeding frenzy occurs inside the mother, where developing young will eat the thickened outer lining of the oviduct. While this might seem like a pretty terrible pre-birth experience, it beats being a mother *Adactylidium* mite (see page 14)—at least your children are only eating *some* of you.

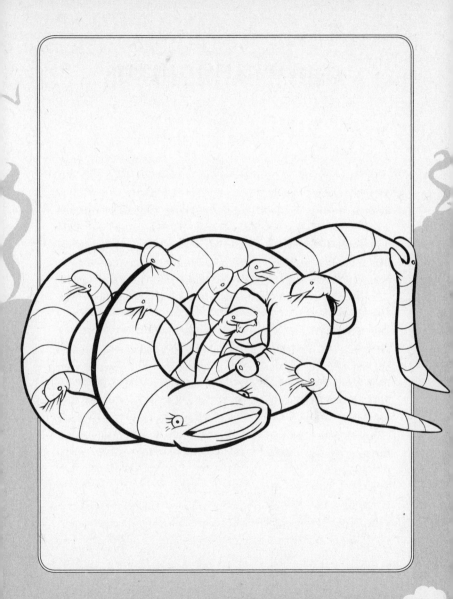

23

ANIMAL EATING HABITS

Animals, as you might imagine, eat a whole variety of foods. What might seem completely unappetizing to you can be a delicious meal for many creatures. You probably already know that some animals are carnivores—that is, they eat other animals; while some are omnivores—they eat both plants and animals; and some are herbivores—they eat only plants. New research, however, has shown that a lot of animals previously believed to be herbivores, such as deer and hippos, are actually more omnivorous than we thought. Deer have been recorded eating baby birds, and hippos will even eat other hippos!

There are also a whole range of other weird and wonderful diets out there—such as planktivores, which feed on (unsurprisingly) plankton; folivores, which eat leaves; coprophagic animals, which feed on poo; hematophagic animals, which feed on blood; and saprophytic animals, which feed on dead organisms. There are lots of myths and legends that revolve around what animals eat; for example, whales, you may not be surprised to hear, don't actually eat people.

If you have read *Does It Fart?* you will have quickly realized that animals don't have a whole lot of manners. Not only does this apply to passing wind, but also to their eating habits. Animals have some pretty bizarre ways of collecting, catching and consuming their food: cats play with their prey before killing it, flies vomit on it, and aye-aye, (a species of lemur from Madagascar) extract their food from trees with long, spindly fingers. This trait has even lead to a common myth on the island that you will be cursed if one of these animals points at you! To date, no wild animal has been recorded using a knife and fork, although some animals, such as New Caledonian crows (a species of crow that lives in New Caledonia, an island off the coast of Australia), will use tools such as sticks to retrieve their food.

In this segment you can find out some gross ways by which animals capture (woodpecker, see page 39; moray eel, see page 32; anteater, see page 34; velvet worm, see page 43) and kill (shrike, see page 40; Komodo dragon, see page 29) their prey, how exactly ticks get to your skin to feed (see page 36), some potentially surprising animal eating habits (moths, see page 28; polar bears, see page 31; vampire bats, see page 44), and drinking habits (camels; see page 26), and some details on animal interactions of both the extant (spiders sleep, see page 48) and extinct (*T. rex*, see page 46) species. Just maybe don't read it while you are eating.

CAMELS STORE WATER IN THEIR HUMPS

TRUE OR POO? POO

There are two species of camel—which is a type of even-toed ungulate in the genus *Camelus*—that are alive today. The dromedary camel (*Camelus dromedarius*) is native to the Middle East and the Horn of Africa, as well as being an invasive species in Australia, and has one hump. The Bactrian camel (*Camelus bactrianus*) is from Central Asia, and has two humps. A common myth about camels' humps is that they use them to store water. (Although it is true that camels can go an incredibly long time without drinking water—over 10 days!—when by comparison a human can live just 3–5 days without water.)

So, if it's not water, what is in that hump? The answer is a whole lot of fat. Camels store fat for energy, as there isn't a huge amount to eat in the desert, but instead of storing it all over their body—which would make them rather warm—it gets stored on their back. Camels can have humps that weigh as much as 36 kg. So how does a camel go so long without water? Well, they can drink a massive amount very quickly, gulping down up to 200 litres in just three minutes, which can keep them going for a while. On top of this they have a whole lot of ways of preventing water loss, including their insulating coat, special nostrils to avoid expelling water when they breathe, and incredibly efficient kidneys that reabsorb most of the water from their urine, making the remainder thick and syrupy. Nice.

ALL MOTHS EAT CLOTHES

TRUE OR POO? POO

Only a very, very, very (very) tiny number of moths eat clothes—four species out of over 160,000, in fact. It isn't even the adult moth itself that eats your clothes, it is actually their larvae—adult clothes moths (genus *Tineola*) have no mouth parts so they do not feed on anything at all. Instead, they do all their eating as a caterpillar—living short lives of a few days once they metamorphose into adults, during which they just manage to mate and lay their eggs. The caterpillars can feed on most natural fibres, including keratin (what our nails are made from), hair, wool, and dust. Before humans came along, clothes moths had to find feathers and shed animal hair to eat—a pretty scarce resource—but we have kindly gathered all their favorite foodstuffs into one place for them in our houses.

Clothes moths are only one tiny proportion of the moth species—just 0.002 per cent. There is a huge array of other, nonclothes-eating moths out there, although identifying them can be tricky. It is harder to distinguish moths from butterflies than you would think, as some moths are brightly colored and some butterflies brown, some moths come out in the day and some butterflies at night. One thing is certain, though, if you see a brown winged insect, moth or butterfly, it is most likely going about its business of eating plants, nectar, or leaf litter, as opposed to your clothes!

KOMODO DRAGONS USE BACTERIA IN SALIVA TO KILL THEIR PREY

TRUE OR POO? POO

For years the method by which the world's biggest lizards, Komodo dragons (*Varanus komodoensis*), killed their prey was thought to be through using bacteria-infested saliva, which was believed to be at least partly caused by a disgusting diet, or the presence of rotting flesh in their mouths and between their teeth. But none of this is true. Scientists now know that Komodo dragons actually *produce* venom that, when injected into their prey via a bite, causes a rapid loss of blood pressure and blood loss by preventing the wound from clotting. Studies have also shown that these lizards don't harbor lethal strains of bacteria in their saliva; so if an animal gets bitten by a Komodo, survives, then gets an infection, this is likely due to their wound not closing in a timely manner and therefore getting an opportunistic infection, rather than it being directly caused by the bite. Surprisingly, Komodo dragons are actually pretty clean eaters, too. They've been observed cleaning their mouths with leaves and they will even swing intestines around their heads to remove the feces before consuming the entrails. With those manners, who *wouldn't* want a Komodo as a dinner guest?

POLAR BEARS EAT PENGUINS

TRUE OR POO? POO

Polar bears (*Ursus maritimus*) are large (up to 700 kg) mammalian carnivores that eat primarily seals but will also scavenge carcasses of whales, eat small rodents or fish, and they are even inclined to rummage through the rubbish. But one prey item that polar bears do not eat, despite many depictions of them doing so, is penguins (family *Spheniscidae*). This is not because they don't like to eat them (they probably would love to!), or because penguins have an evolutionary adaptation to deter predation by polar bears, it is simply because polar bears and penguins do not live in the same place. Polar bears are native to the Arctic Circle in the Northern Hemisphere, whereas penguins are almost entirely native to the Southern Hemisphere (except for the Galapagos penguin, *Spheniscus mendiculus*, whose range just crosses the Equator into the Northern Hemisphere). In fact, the word "Arctic" originates from the Greek word *arktos*, which means bear, and Antarctica, which is perhaps where most people are familiar with penguins such as the emperor penguin (*Aptenodytes forsteri*), is defined as opposite (Ant-) of bear (-arctica): so there is *literally* nowhere less likely to feature polar bears than Antarctica.

MORAY EELS HAVE MORE THAN ONE SET OF JAWS

TRUE OR POO? TRUE

There are over 200 species of moray eel on this planet, ranging from the tiny Snyder's moray (*Anarchias leucurus*) at just 11.5 centimeters, to the massive, and very aptly named, giant moray (*Gymnothorax javanicus*), which reaches over 3 meters in length and can weigh up to 30 kg. They mostly live in marine environments, and feed primarily on fish and shellfish. They have one particularly cool trait that is common across all moray species: a second set of jaws. These mini jaws, the pharyngeal jaws, are found in the moray's throat. As opposed to other fish, which suction up their food, moray eels bite their prey before this second set of jaws shoots forward, grabbing the food and pulling it into the eel's throat. This is because, in order to fit into the rocky burrows in which most moray species live, their heads are too narrow to create the suction needed to suck food into their throat. While these jaws may seem terrifying, and reminiscent of an alien, you needn't worry; these eels may look scary but they will only bite if you disturb their burrow, or if you decide it is a good idea to feed them, which can lead them to mistake your hand for food. So it's probably best to not stick your hands in any strange holes at the bottom of the sea.

ANTEATERS USE THEIR NOSES TO SUCK UP ANTS

TRUE OR POO? POO

There are four species of anteater alive today; two species of tamandua (ginger and black tree living anteaters), the silky anteater (*Cyclopes didactylus*) and, probably the most well-known, the giant anteater (*Myrmecophaga tridactyla*): but contrary to popular belief, they're not some sort of ant-vacuum. The anteater's long snout is actually both their mouth and their nose combined. If an anteater were to use their nose to suck up ants, a) their food would get covered in snot (and nobody wants that), and b) the ants or termites would bite and sting the inside of their nasal canals, which would be very uncomfortable. Instead, anteaters use their incredibly long tongues—up to 60 centimeters long in the giant anteater—to lap up ants into their mouths before swallowing them. They can flick their tongues up to 160 times per minute, and these tongues are covered in thousands of tiny hooks, called filiform papillae, which hold the ants onto the tongue with the aid of a lot of sticky saliva. Anteater mouths are so specialized for eating ants as quickly as possible that they evolved to have very limited jaw movement and have no teeth. Instead, they crush the insects against their palate before swallowing. These incredible adaptations allow the giant anteater to consume over 20,000 insects a day!

TICKS CLIMB TREES AND DROP ON YOU FROM ABOVE

TRUE OR POO? POO

It is pretty common for people to find ticks on the top part of their body, including their upper arms, shoulders, or even head. It would seem obvious that these little critters dropped down out of the trees onto their unsuspecting victim, but in fact ticks can't jump, and much more commonly they hang out at between ankle and waist height—where their usual hosts, such as sheep, deer, or racoons, are most likely to brush past them. So how do ticks get onto our upper bodies, and why do they do it? Well, ticks, which are in fact not insects but part of the arachnid family (along with spiders, (see page 48) and scorpions (see page 70), have a propensity to climb. To find a suitable host, ticks do what is known as questing, where they climb onto high foliage and hang out in waiting. They use something called a Haller's organ to detect carbon dioxide and body heat from nearby animals. When an animal (or human!) brushes past, they will grip on with their front legs. Often when they grab onto a human they encounter clothing instead of bare skin. So they do what comes naturally—they continue to climb until they find a suitable feeding spot. Of course, once they do, they latch on with gusto, sucking your blood until they have ingested up to 400 times their own weight!

WOODPECKERS WRAP THEIR TONGUES AROUND THEIR SKULLS

TRUE OR POO? TRUE

If you thought anteater (see page 34) tongues were weird, wait until you hear about the freakishly long tongue of the woodpecker, which wraps around their entire skull. Woodpeckers are members of the family *Picidae*, and, unsurprisingly, they all peck wood. The distinctive hammering for which they are famous is generally done to dig out insects hiding out inside trees and rotting wood. (Some species, such as the acorn woodpecker, feed on seeds, and the pecking is instead used to create storage space for their food.) Woodpeckers have particularly long, bristled tongues, which many species use to delve into the holes they have created and grab their prey. In woodpecker tongues, instead of mostly being made up of muscle, like in humans, the bony structure called the hyoid extends through the tongue's entire length. In some species their tongue even goes all the way along the beak, into the head, around the back of the skull and ends in the nasal cavity! This unique bony tongue helps protect the woodpecker's skull from all the high-impact knocks it gets, acting like a crash helmet—otherwise they would get concussion every time they ate.

SHRIKES KILL THEIR PREY ON FENCES

TRUE OR POO? TRUE

Ah, the countryside—beautiful rolling hills, meadows of flowers, and... dead lizards? If you've ever come across a lizard or other animal impaled on a barbed-wire fence, this is most likely the work of a shrike (family *Laniidae*). These birds use sharp objects such as thorns, barbs, or even the prongs of a garden fork to impale their prey, which consists of reptiles, rodents, smaller birds, and insects. This seemingly ruthless behavior allows shrikes to keep their meal pinned down so that they can more easily tear into their flesh. Shrikes will continue to impale prey even if they aren't currently hungry, saving their butchered meals for leaner times, with the added bonus that decomposing flesh is easier to tear. Spearing prey is essential to shrikes; some species have been observed to lance nonanimal food such as dates, and captive birds have refused food until they were provided with a sharp object. One individual even needed to spear a piece of cheese before eating it! But that's not all, male shrikes also use their butchered food storage to attract females to their territory, and will skewer inedible and colorful objects such as snail shells and ribbons to boost the size of their cache.

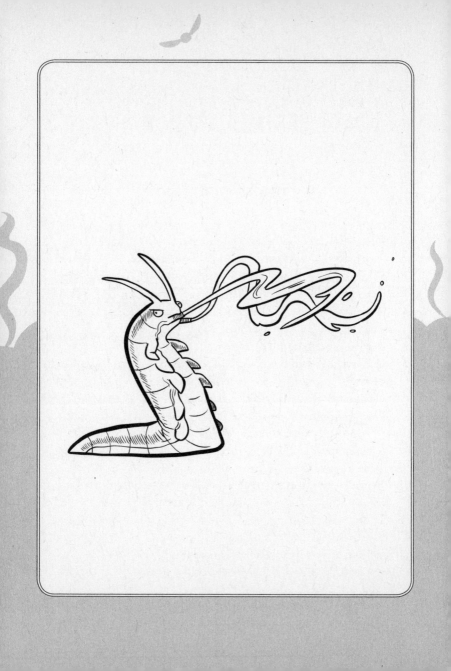

VELVET WORMS IMMOBILIZE THEIR PREY BY TURNING THEIR FACES INTO SLIME CANNONS

TRUE OR POO? TRUE

Hunting with a poor sense of vision seems like a significant disadvantage, but the velvet worm has a unique evolutionary adaptation to overcome it: the slime cannon. Velvet worms (phylum *Onychophora*) are segmented, soft-bodied arthropods that feature numerous legs and, true to their namesake, a velvety appearance that is caused by tiny sensory bristles protruding from papillae, or small round bumps, all over their body. Velvet worms detect prey using the changes in the air currents that are caused by the prey's movement, then they will shoot slime from their oral papillae, which are located just below their antennae on either side of their head. Their slime cannons act like two untethered garden hoses on full power, swaying back and forth while shooting a glue-like slime up to a foot from their body. Any prey unlucky enough to get slimed is quickly ensnared as the goo hardens, leaving the prey animal unable to break free. This means that the velvet worm can walk right up to its dinner, puncture its exoskeleton with its sharp jaws and inject their saliva into it, which begins the process of digestion. They will also salivate on their slime and ingest that as well. Waste not, want not.... Not surprisingly, the ability to shoot glue is a pretty good defence, too, and velvet worms will turn their slime cannons on potential predators if they need to escape danger.

VAMPIRE BATS SHARE FOOD WITH THEIR FRIENDS

TRUE OR POO? TRUE

When you hear "vampire bat" do you think of a terrifying blood-feeding menace? Or do you think cute, fluffy, sky puppy that takes great care of its friends? Well, if you went with option one, you may be surprised. There are three species of bat found in South and Central America that feed exclusively on blood—biting their prey while the animal sleeps using specialized teeth and producing an antiseptic, anticoagulant saliva that prevents the blood clotting while they feed. While this has earned these bats a pretty horrifying reputation, they are actually incredibly social creatures, living in colonies of hundreds of individuals, and forming strong bonds with their colony mates. Female bats will groom each other and even share food. If a bat hasn't managed to get a meal in a previous night, one of its neighbors may share their meal with it, then in the future, if that neighbor goes hungry, the bat that was previously on the receiving end of the blood-sharing will return the favor. This is known as behavioral reciprocity, whereby animals that are not related increase their chances of survival by sharing resources. This trait is relatively rare in nature, making the vampire bat a particularly altruistic species.

A *T. REX* CAN'T SEE YOU IF YOU STAND STILL

TRUE OR POO? POO

In probably the most famous dinosaur-related film of all time, the protagonists, upon being attacked by a *Tyrannosaurus rex*, are instructed not to move, because if they hold still the dinosaur will not be able to spot them. In reality, however, it turns out that *T. rex* probably had pretty good vision. It may even have been better than modern-day raptors—that is, eagles and hawks—which are species famed for having incredibly good vision. Not only that, but *T. rex* was also equipped with a great sense of smell. Remaining immobile doesn't do much to mask your scent, so even if the *T. rex*'s vision was poor, standing still probably wouldn't do you a whole lot of good. In fact, contrary to what we often see in films, standing still and staying quiet is a pretty bad bet when faced with any large predator (as the screaming hairy armadillo, see page 106, well knows). Top predators have evolved to have great eyesight, smell, and hearing in order to better locate prey, and so will be able to locate you with ease. Life finds a way, after all.

YOU EAT SPIDERS IN YOUR SLEEP

TRUE OR POO? POO

While there are many myths about animals eating people, there are also a fair few about humans eating animals! Many species of spiders are nocturnal, which means while you are sleeping they are out hunting, and some species may do their hunting inside your home. So does that mean that if you sleep with your mouth open a spider will crawl in and you'll eat it? While it would be nearly impossible to confirm unconscious nocturnal arachnid consumption, you can rest easy tonight (as well as every other night) that you will probably not be eating a spider in your sleep. This doesn't mean that you've never eaten a spider in your sleep (a few of you probably have), it's just not a regular occurrence. Spiders aren't really interested in going near your mouth. Would you be interested in going near something's mouth that is 100 times larger than you, makes a lot of noise and vibrations (snoring, breathing, and your heart beating), and is moving?

The more likely scenario, however, is that you've probably eaten a spider, or maybe another insect, in your food. For example, the U.S. Food and Drug Administration reports that there could be as many as 400 insect fragments per 50 grams of ground cinnamon, or 10 or more whole (or equivalent) insects as well as 35 fly eggs per 225 grams of golden raisins. So it seems that the only time you *aren't* eating spiders is while you are sleeping. Sweet dreams!

DIGESTION AND EXCRETION

In between munching down a meal and/or quaffing a drink and pooing/weeing/sweating out/vomiting up the leftovers there are a whole variety of digestive processes going on, involving a whole host of specialized organs (depending on the animal). When humans eat, the food goes through the mouth, esophagus, stomach, small intestine, large intestine, and rectum before being excreted through the anus. This is the same for most mammals. Invertebrates generally have more simple digestive systems; for example, jellyfish take food in through their mouths into a single cavity where it stays until all the nutrients are absorbed. Waste is then expelled back out through the mouth. Insects have a few more organs that we recognize from our own digestive systems: a foregut, a midgut, a hind gut, and a rectum. Most birds have a specialized organ known as the crop, which temporarily stores food that they regurgitate for their offspring, and a gizzard, which often contains stones that the bird has ingested which help to grind up hard food, as birds do not have teeth.

What goes in, must come out—and this is true for the entirety of the animal kingdom! Digestion is the process of breaking down food, and excretion is how living organisms get rid of the waste. Animals excrete their waste products in many different ways—whether it be defecation and urination, as in humans and other mammals;

a combination of the two, as we see in birds and reptiles; the regurgitation of indigestible parts, as we see in birds of prey; or just simply excreting it out through their skin, as seen with ammonia in many aquatic animals. In fact, even humans excrete a waste product from digestion—urea—through their skin when they sweat. The products of excretion come in many forms and lead to some weird and wild animal observations.

In this section you will learn about animal digestive anatomy (mammal stomachs, see page 52), find out alternative uses for the digestive system (gastric frogs, see page 54), discover whether microbes can help as well as harm (microbes, see page 53), if all animals can vomit (horses, see page 66), and find out what beaver butts have to do with vanilla ice cream (beavers, see page 69). You will also discover some of the wonderful things that poo can do for both animals (vulture, see page 61; termites, see page 63) and people (parrotfish, see page 56), explore how poo can help researchers (penguins, see page 58), learn about the frequency with which bees defecate (bees, see page 59), discover some interesting animal poo shapes (wombats, see page 62) and habits (shrew pitcher plants, see page 64; rabbits, see page 68), and find out when the inability to poo can be deadly (scorpion, see page 70).

ALL MAMMALS HAVE STOMACHS

TRUE OR POO? POO

If you thought stomachs were a necessary part of mammal digestive systems, you would be wrong! You probably already realized that platypuses (*Ornithorhynchus anatinus*) are pretty weird. Number one, they are a mammal that lays eggs. Two, they have a beak. Three, they are venomous. Yes, that's right, the cuddly, furry, adorable platypus—or at least the male of the species—has venomous spurs located on their hind limbs. Less well-known is the fact that they don't have stomachs. In fact, no monotremes (the egg-laying mammals that also include echidnas) have stomachs. A stomach is the part of the digestive system where acids and pepsins are produced by gastric glands to help digest food. In platypuses, this is absent— the esophagus connects straight to the hind gut. It is thought that while platypus ancestors did have a stomach, it was lost, potentially because their diets did not need the aid of pepsins—which only work in acidic conditions—to digest their food, and therefore over time they evolved to stop producing acid. Alternatively, many animals that have lost their stomachs have diets high in chalky or carboniferous items, such as the bumphead parrotfish (see page 56), meaning the acid in the stomach would be neutralized, rendering its production pointless.

ALL MICROBES ARE BAD AND CAN MAKE YOU SICK

TRUE OR POO? POO

Walk down the cleaning supplies aisle at the supermarket and you might see claims like "kills 99.9 per cent of bacteria/germs," but is this necessary, or even a good thing? Microorganisms, or microbes for short, are living organisms that cannot be seen without the aid of a microscope and include bacteria, fungi, protists, or archaea. Microbes can be found everywhere, and some can infect human cells, causing illnesses like giardia or malaria. But microbes are also found everywhere on the human body, and they are numerous; in fact, scientists estimate that for every one human cell, the average human contains 10 microbe cells, and the number of microbe genes in the average human body may outnumber the human genome by up to 1000 to 1! These microbes are essential to our health—they play a key role in our immune, digestive, and nervous systems. For example, studies have shown that the removal of portions of our microbiome, through use of antibiotics, can lead to a vitamin K deficiency. Unfortunately, through the overuse of antibiotics humans have also caused the evolution of antibiotic-resistant strains of bacteria, which can be harmful or even fatal. Obviously, we do not suggest that our readers never wash their hands, bathe, clean, or use antibiotics as prescribed; however (and we apologize if this is used by kids trying to get out of their chores), there is such a thing as too clean!

SOME FROGS RAISE THEIR YOUNG IN THEIR STOMACHS

TRUE OR POO? TRUE—BUT NOT ANY MORE

It turns out that inside their back (see page 19) isn't even the strangest place that frogs are known to raise their young. Gastric-brooding frogs, or platypus frogs, are two species of frog from Queensland, Australia. These frogs have a very unique method of reproduction—once the female lays her eggs and they are fertilized by the male she puts them in her mouth and swallows them. This may seem like terrible parenting—why is she eating her own children?!—but fear not, this is all part of the plan. The eggs contain a special chemical, called prostaglandin, that stops the female's stomach producing hydrochloric acid, which would otherwise cause the eggs to be dissolved and digested. Once the tadpoles hatch from the eggs they also produce prostaglandin, and continue to grow and develop inside their mother's stomach. During the development process the mother cannot eat, and her stomach expands to fill most of her body, making up over 40 percent of her body weight. As the tadpoles mature into froglets they are regurgitated one at a time from the mother's mouth—she vomits up her own children! Sadly, both species of gastric brooding frog went extinct in the 1980s due to an introduced fungus, so this unique reproductive cycle is no longer observed in nature.

WHITE SAND IS MADE OF FISH POO

TRUE OR POO? TRUE (SORT OF)

While the majority of sand is mostly poo-free—you know, bar the poo of the crabs, worms, and other organisms that live in it—the famous white beaches of the Maldives are basically just one giant beautiful accumulation of fish poo. Or at least, a component part of fish poo. Not just any old fish, though; this sand is formed by one fish species in particular, the bumphead parrotfish, *Bolbometopon muricatum*. This large green fish feeds almost entirely on coral (see page 113). Coral contains a type of algae, the foodstuff of most parrotfish species. Bumphead parrotfish bite off bits of coral with their hard beaks, which continuously grow throughout their lifetimes, before chewing it into dust with specialized pharyngeal teeth, digesting the algae, and leaving behind the coral's powdered, rocky remains. This ground-up coral is then pooed out as a fine sand, before being washed ashore to form the stunning beaches for which the Maldives are famous. Each parrotfish can produce 90 kg of sand each year, and as a result 85 percent of sand in the region has passed through a parrotfish at some point. So if you are ever lucky enough to visit the Maldives, you can relax on the beach knowing it all would not have been possible without one special parrotfish's sandy stools.

YOU CAN SEE PENGUIN POO FROM SPACE

TRUE OR POO? TRUE

Penguin poo—pretty cool, right? This fishy, smelly substance has been invaluable in discovering the secrets of the species. Back in 2009 advances in satellite technology allowed researchers from the British Antarctic Survey to scan the Antarctic ice flows for large brown patches… yup, giant piles of penguin poo. These scientists discovered this purely by accident when checking out one of their study colonies on satellite imagery; they realize that icy poo stains were a telltale sign of an emperor penguin (*Aptenodytes forsteri*) colony. This led to the discovery of ten more penguin colonies, containing tens of thousand of individuals. Since then, scientists have used satellites to discover nearly 50 percent more emperor penguin colonies over the course of just a few years, nearly doubling the known population. In more recent years, satellite imagery has advanced enough that it is even possible to estimate colony density—i.e., the number of birds in each colony!

This isn't the only awesome data that penguin poo has uncovered, by taking sediment cores and observing the amount of penguin poo that was deposited each year, a team of scientists were able to discover that a colony of gentoo penguins (*Pygoscelis papua*) on Ardley Island, Antarctica, was nearly wiped out three times in the past by volcanic activity—a penguin Pompeii, if you like.

BEE LARVAE ONLY DEFECATE ONCE

TRUE OR POO? TRUE

Workers of the honey bee (genus *Apis*) don't defecate inside their hives, for obvious reasons, but rather they will hold their waste until they are flying away from the hive and drop their loads mid-flight. In fact, if bees have been cooped up in their hive for extended periods of time due to cold weather, fecal trails near the hive may be seen soon after temperatures become warmer. But larvae lack the ability to fly and must stay within the confines of the hive, which means their poop must stay, too. Honey bee larvae eat a lot and grow rapidly, they can have as many as 1,600 total feedings and increase their weight by 1,700 times! Since what goes in must come out, that could be a very messy hive. Fortunately, larvae only defecate once. But it's not that these larvae don't want to poop, they simply can't poop. During the majority of the larval stage, two portions of the bee's digestive tract, the hind gut and midgut, aren't connected, so the feces have nowhere to go. It isn't until shortly before their pupal stage—that is the stage between the larval and mature stages while encased in a cocoon—that these two sections of their digestive tract are joined and the larvae can and will finally expel their waste. Fortunately (unless you happen to be a nurse bee) nurse bees, or younger worker bees, will clean up this one-time mess to prepare the hive for the next batch of larvae.

VULTURE POO KILLS BACTERIA

TRUE OR POO? POO

Vultures poo on their own legs. A lot. Vulture poo, like all bird poo, is made up of both feces and urine, and is therefore liquid. One common myth is that they poo on their legs because it kills bacteria that they pick up from standing around dead carcasses all day. This isn't true—in fact, vulture poo is pretty jam-packed with bacteria, including those that make most animals very sick. Bacteria found in vulture poo even includes *Clostridium*—a family that includes bacteria that causes botulism and tetanus, and it is thought that these bacteria actually help the vulture digest its food. Vultures have incredibly tough digestive systems to enable them to eat carrion, so those micro-organisms that do survive are particularly potent. Don't worry, though, you won't get sick from a vulture, unless you literally eat its poo (disclaimer: definitely DO NOT eat vulture poo), in fact, vultures help prevent disease, including anthrax, by removing carrion from the environment.

So why do vultures poop on their legs? Well, as vultures can't sweat, they need another way to cool off. Their legs are full of blood vessels close to the surface of their skin, so as the liquid poo evaporates it helps them lose heat and keeps them cool.

WOMBATS HAVE CUBE POO

TRUE OR POO? TRUE

Like the honey bee (see page 59) this is one poo that is true. Anyone who played with modelling clay as a kid knows that you can create shapes in the clay by pushing it through a hole that has the desired shape. But does that mean that the wombat's (family *Vombatidae*) cube-shaped poo is caused by a cube-shaped rear end? Unfortunately (or fortunately, depending on your comfort level with poo), this is not the case. The wombat's herbivorous diet takes a relatively long time to pass through its digestive tract—up to 18 days!—which means there's a lot of time for the feces to become dry and compacted, as water is continuously reabsorbed. The feces are likely molded in the beginning of the large intestines, known as the proximal colon, due to the colon's ridges, and is able to retain its shape as it passes through because the end of the large intestine, the distal colon, is smooth. Perhaps the most fascinating part of the wombat's poo, however, is that it is an evolutionary adaptation! The wombat's sense of smell is better than their eyesight and they use their feces to mark their territory on top of objects such as rocks or logs, known as markers. Wombats can produce up to 100 of these scat markers a day; cube-shaped poo is less likely to roll off wherever it's left, so the more cuboid the poo, the better the wombat is at marking its territory.

TERMITES USE POO TO HELP THEM BREATHE

TRUE OR POO? TRUE

While using your poo to keep cool seems pretty foul, termites use it for something that may sound even more disgusting. Mound-building termites are impressive animals—colonies of some species can move over a ton of dirt to construct their homes, which can be over 6 meters high! Termites don't actually live in their mounds; the colony, which can contain millions of individual termites, lives underneath it. The structure, with its complicated tunnels and passageways, facilitates gas exchange to the colony below, allowing the termites to breathe. Termite poo, known as frass, can play a surprisingly important part in this. Some species of termites, such as the harvester termite, *Microhodotermes viator*, build most of their mounds out of poo, gluing it together with their sticky saliva. In some species it has been shown that building mounds with poo encourages the growth of bacteria that have anti-microbial properties, killing pathogens that would otherwise make the colony sick. This isn't the only important role that poo plays in termites' lives, though. The largest chamber under the mound is known as the fungus garden, and in here termites build structures from their own poo, seeding them with a fungus. This fungus helps to break down the wood and cellulose in the feces, making the nutrients digestible for the termites.

SHREWS USE PITCHER PLANTS AS TOILETS

TRUE OR POO? TRUE

Carnivorous plants that have modified leaves that form pitfall traps to trap and kill prey are collectively known as pitcher plants—a group of plants that possess a cupped leaf containing digestive enzymes underneath a lure, which they use to attract animals, most often insects, which are trapped and then digested. Some pitcher plants, however, evolved not to trap living prey but rather to obtain their necessary nitrogen by providing an ideal bathroom break. Scientists have found that three species of pitcher plant in the genus *Nepenthes* act as both a feeding station and toilet for the mountain tree shrew (*Tupaia montana*). Unlike other pitchers, these plants don't have a slippery rim: instead, the distance between the pitcher opening and lid matches the shrew's body length such that the shrew's rear end is well-positioned over the pitcher orifice, or opening. Additionally, the plants have a high amount of nectar, and are more reinforced, letting them take the weight of the visiting shrew. Because who would want to use a bathroom that tips over? Tree shrews will even mark their bathroom/feeding stations by rubbing their genitals on them, ensuring they can return to the same plants. We must insist, however, that our readers do not follow the tree shrew's example; many species of pitcher plants are threatened with extinction and can be very fragile, so please don't poo in them (or rub your genitals on them, for that matter).

HORSES CAN'T VOMIT

TRUE OR POO? TRUE

Puke, chunder, spew, throw up, barf, hurl, ralph, chuck up—all delightful slang alternatives for the medical term "to vomit." When we vomit our stomach is squeezed between our abdomen and diaphragm, and the food and other contents are expelled. This generally happens because our body doesn't want us to get poisoned, or because our stomach is irritated. Horses, however, can't do this. The valve at the bottom of the esophagus, which leads into the stomach, is much stronger in horses than humans, preventing food being passed back into the mouth. On top of this, a horse's stomach is positioned within the rib cage, preventing it from being squeezed by the abdominal muscles, and horses have a low vomit reflex—the neural pathway that tells an animal they need to heave. Vomiting is a pretty unpleasant process on the whole, so you may be thinking "lucky horses!" Sadly, however, this can cause all sorts of problems, because if horses eat the wrong food, such as grass clippings, they can become painfully bloated—known as colic. This causes serious health problems and can even be fatal. The inability to vomit is a rare trait among mammals, but horses share it with rodents, which have instead evolved to be incredibly adept at tasting toxicity in their food.

RABBITS EAT THEIR OWN POO

TRUE OR POO? TRUE

Yes, that's right, cute, adorable bunnies eat poo. Why, though? Rabbits are hind-gut fermenters, meaning their food is broken down in the caecum, right at the end of their digestive system. Nutrients, however, are absorbed in the rabbit's stomach and small intestine, through which the food has already passed. For rabbits, the solution is to produce a special, nutrient-rich type of poo—called ceaocotrophs—from their caecum, which they then re-ingest in order to be able to absorb the nutrients from their food. Rabbits are far from the only animals that eat poo, though—a behavior known as copraphagy. Birds (see page 20), dogs, guinea pigs, koalas, pandas and elephants are all regular poo eaters. As you can imagine, guinea pigs eat their own poo for similar reasons to rabbits. In koalas, pandas, and elephants, it is essential that the offspring ingest their mother's poo—they have to do this to obtain important gut flora from their parent, which is critical for the digestion of their plant-based diets. Baby elephants have even been observed reaching into their mother's bum to pull out a sloppy mouthful. Dogs, which are carnivores, are a bit of an outlier, as they don't gain an obvious nutritional benefit from this act. They aren't just being gross for the sake of it, though; if your dog is eating poo it can be a sign that something is missing from their diet.

BEAVER BUTT GLANDS ARE IN YOUR FOOD

TRUE OR POO? MOSTLY POO

Beavers (genus *Castor*) produce a compound known as castoreum, a yellowish-brown secretion from their castor glands, which they use for marking their territory and general scent communication. Though this scent originates from glands that can be more accurately described as anus-adjacent, it has a rather pleasant, vanilla-like odor, which comes from the beaver's diet of leaves and bark. Humans have a long history of using castoreum: it was used to alleviate headaches (it does contain salicylic acid, the main ingredient in aspirin) and treat anxiety, it has been used in perfumes, and, yes, it is used as a food additive. But don't start throwing out all food with this ingredient that has a vanilla flavoring. Milking beaver secretions to extract castoreum is too expensive to enable it to be manufactured on a large scale—only around 45 kg are consumed each year, compared to vanilla flavor extracted from vanilla beans, of which approximately 900,000 kg are consumed each year. So while the extract from a beaver's castor gland can be in food (and is perfectly safe to be there and for you to digest!), it probably isn't in yours. Probably...

SCORPIONS CAN DIE OF CONSTIPATION IF THEY SHED THEIR TAILS

TRUE OR POO? TRUE

Tail autotomy, or the discarding of a tail when an animal is threatened, is probably best known in lizards; however, recent studies have shown this anti-predation behavior in 14 species of scorpions in the genus *Ananteris*. In lizards, the cost of losing a tail can include losing fat reserves, a decrease in speed, reduced reproductive potential, or maybe even lowered social status (although these costs are much preferred to death!). However, when a scorpion autotomizes its tail to escape from a predator, its maximum life expectancy becomes around eight months. This is because scorpions, unlike lizards, lack the ability to regenerate their tails. A scorpion's tail contains two crucial organs: their stinger and their anus. This means their ability to capture prey or defend themselves is compromised but, most importantly, they lose their ability to poo. Therefore, waste will continue to build up, and in some cases the abdomen will swell and additional tail segments may even break off. This might sound unpleasant or undesired, but male scorpions don't live long at the best of times and may only have one opportunity to mate in their lifetime. So enduring a fatal bout of constipation is more than worth it.

DEFENSE

Most animals need to protect themselves from attack because pretty much everything around them will try to eat them if they let their guard down. The best defence is simply to be completely hidden from predators at all times, but this isn't always possible. Animals need to find food (see animal eating habits, page 24) and mates (see courtship, mating, and parenting, page 1) and often these basal needs increase their risk of being something else's next meal. Many animals seem to live by the adage "the best defense is a good offense" and will inflict pain on their attacker to discourage attacks. Humans might think that a particular animal or species is "aggressive," but the more likely scenario is that we have encroached on their territory and their aggression is an artifact of them feeling cornered. This is why wild animals are always best viewed from a safe distance, even if it seems as if they are docile.

Animal species have evolved some fascinating adaptations to keep their predators away from them. For example, some cobra species (genus *Naja*) will spit their toxic venom, which can cause blindness, as far as 2 meters! Tail autotomy might be the best-known method for escaping a predator whereby an animal (such as lizards, salamanders, or even some mammals) will discard their tail, which can regenerate, when grabbed by a predator. But some species can discard other appendages or even organs that can still be regenerated! The worst-case scenario for animal is surely when they find themselves in a predator's jaws. However, they can still keep fighting by producing toxins that are distasteful or harmful or that would otherwise prevent predators ingesting them.

In this section you learn the answer to questions about some of the most disgusting ways that animals have evolved to stay alive, and have a few myths busted about animal defense along the way. What is the true reason for the chameleon's colors? (See page 74.) Are younger snakes truly more deadly than their older counterparts? (See page 75.) Do wasps have a secret grudge against humans? (See page 90.) What exactly is a daddy longlegs, and can it harm you? (See page 80.) Which species use projectiles to defend themselves, and are any of them deadly? (See pages 43, 76, 79, and 83.) Do some lizards autotomize more than just their tails? (See page 70.) Can earthworms regenerate their whole bodies? (See page 87.) Do toads give you warts to discourage picking them up (See page 89), or is there something more sinister lurking in their skin? (See page 91.) Do any animals harm themselves with their defense strategy? (See page 92.) And are sharks truly invulnerable to cancer? (See page 81.)

CHAMELEONS ARE MASTERS OF CAMOUFLAGE

TRUE OR POO? POO

Chameleons (family *Chamaeleonidae*) have a number of fascinating adaptations: their eyes can rotate 180 degrees, their tongue, which is the largest in proportion to its body size in the animal kingdom (about twice as long as their body), can be propelled up to 26 body lengths per second (that's about 13 miles an hour), and they have a sticky mucus secretion to adhere their tongue to prey items that's around 400 times more viscous, or thicker, than human saliva! But perhaps most fascinating, and the characteristic for which chameleons are best known, is their ability to change color.

Chameleons are able to change color using an outer layer of pigmented skin: not, though, because they want to blend into their surroundings. Instead their skin color reflects their mood or temperature. The most superficial, or outermost, layer of skin can change the arrangement of nanocrystals within iridophore, or reflective, cells. When the chameleon becomes excited (like when a male sees a rival male), the nanocrystals are further apart, reflecting longer wavelengths of light (such as orange or red), whereas a relaxed chameleon will have nanocrystals that are closer together, which reflects shorter wavelengths, like blue.

BABY SNAKES ARE MORE DANGEROUS THAN ADULT SNAKES

TRUE OR POO? POO

If you have ever been out and about somewhere where there are venomous snakes, you may have heard something along the lines of young snakes being more dangerous than the adults because they cannot control the amount of venom they deliver. It is so ubiquitous, even appearing on a number of natural history programmes, that you probably think this is true. It is, however, one big myth! In fact, the scientific evidence is not conclusive over whether any snake, young or old, can control the amount of venom it injects—it may just be a case that the amount is dependent on the density of what they are biting. Because smaller snakes contain a lot less venom than larger snakes, realistically even a partial hit from a larger snake is likely to contain a lot more venom than a bite from a small one. In a number of snake species, including Australian brown snakes (genus *Pseudonaja*), their venom actually gets more potent as they get older.

Snakes (like wasps, see page 90) aren't out to get you, though. They will only bite if they feel threatened, so treat all snakes, big or small, with respect and you have nothing to fear.

PORCUPINES SHOOT THEIR QUILLS

TRUE OR POO? POO

There are two families of porcupines, which can be separated into the New World porcupines (family *Erethizontidae*), which are found in North and South America, and the Old World porcupines (family *Hystricidae*), which can be found in Europe, Africa, and Asia. Porcupines are probably most famously known for their quills, which are modified hairs that have a coating of thick plates of keratin and can be up to 35 centimeters long; an individual can be covered by as many as 30,000 quills! While porcupines use their quills for defense, we can use them to distinguish the two families; New World porcupine quills have barbs at their ends and are interlaced with hair along their body, whereas Old World porcupines' quills do not have barbs and are more clustered together. But if you thought that porcupines could shoot their quills, you aren't alone: this myth was infamously spread by Aristotle and has persisted for nearly 2,000 years. When threatened, porcupines will raise their quills, which can be easily detached from their body when touched, and some quills may even fall out when an individual shakes. It seems likely that the ease with which the quills detach from the porcupine's body is what started this myth. However, even though porcupines can't shoot their quills, it's probably still a good idea to keep a respectful distance.

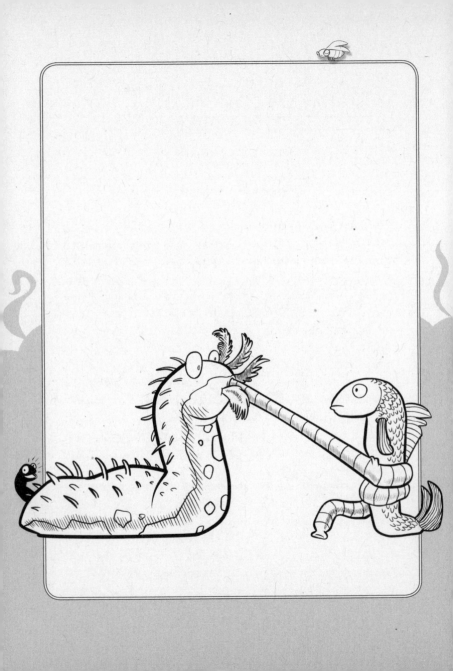

SEA CUCUMBERS THROW THEIR INTESTINES AT PREDATORS

TRUE OR POO? TRUE

Picture this: You are a cucumber-shaped marine invertebrate of the class *Holothuroidea*, aptly named a sea cucumber. You are grazing on mud and other detritus (yum!), minding your own business. A predatory sea star comes along and starts nibbling on you. What do you do? You are super slow, what with being a sea cucumber, so you can't outrun your predator. The answer, of course, is that you expel your guts through your body wall onto your attacker. Or maybe your respiratory tree (your equivalent of lungs). Or maybe your gonads instead—it depends what species of sea cucumber you are.

Sea cucumbers have special connective tissue which can soften or harden, and this can help them anchor themselves in place in rough seas, or even become liquid. When threatened, the connective tissue holding the sea cucumber's organs in place rapidly softens, parts of the body wall become soft, and muscles in the body wall contract—literally shooting the internal organs at the predator. This may seem like a ridiculous method of escape: surely they just die without their internal organs? But, fear not, sea cucumbers have amazing powers of regeneration and will regrow their organs.

In a slightly less traumatic method of defense, some species even have specialized structures called cuvierian tubules—sticky strands that are expelled through the anus to entangle predators.

DADDY LONGLEGS SPIDERS ARE EXTREMELY DEADLY BUT CAN'T BITE YOU

TRUE OR POO? POO

This common myth is a two-for-one poo. The name 'Daddy long-legs', or 'Granddaddy longlegs' (though we aren't sure how people are able to tell if an individual has sired one or two consecutive living generations), may be applied to either cellar spiders (family *Pholcidae*) or, more commonly, harvestmen (order *Opiliones*). While cellar spiders do have fangs and are capable of biting humans (their fangs are short but of a similar size to brown recluse spiders, *Loxosceles reclusa*), their venom is simply not dangerous to us. Harvestmen, on the other hand, are not spiders (although they are arachnids). Harvestmen do not have a separate abdomen or spin silk, which are both characteristics of true spiders. They are also not harmful to humans, as they lack venom glands, and their mouthparts, known as chelicerae, are more analogous to grasping claws than fangs and not strong enough to pierce human skin. Even worse, the name "daddy longlegs" further highlights an issue with common, or colloquial, names, as this term is *also* used to refer to crane flies (family *Tipulidae*), which are not arachnids at all (and are also not venomous).

SHARKS DON'T GET CANCER

TRUE OR POO? POO

The idea that sharks (superorder *Selachimorpha*) don't get cancer is often cited as proof that they are perfect, almost invulnerable predators. Sharks, like other species in the class *Chondrichthyes* (meaning cartilaginous fish), have skeletons made of cartilage rather than bone. It is true that scientists have found that cartilage cells produce a protein that can inhibit the growth of blood vessels (cartilage is avascular—that is, it does not contain blood vessels). One of the ways to treat cancer is to prevent tumors from growing new blood vessels, which deprives it of the ability to receive oxygen and nutrients or get rid of waste. This does not mean that sharks do not get cancer, however. They do—and they can even get cartilage cancer, or chondromas. Even though shark cancer rates have been reported as relatively low, this is more likely due to a shortage of studies that focus on shark cancer rates rather an innate immunity or resistance. Unfortunately, this myth has itself become harmful to sharks, which are harvested for their cartilage in the incorrect belief that it cures or prevents cancer when eaten.

BOMBARDIER BEETLES SHOOT ACID FROM THEIR BUTTS

TRUE OR POO? TRUE

In the book *Does It Fart?* we discussed the beaded lacewing's deadly farts, but it isn't the only insect featuring a fart arsenal (or fartsenal, if you will). The bombardier beetle (family *Carabidae*) stores both hydroquinone and hydrogen peroxide and, when disturbed, mixes them together within another chamber in its abdomen. This caustic solution is ejected along with oxygen gas in pulses (up to 735 a second!) from pygidial glands at the posterior, or end, of their abdomen. Moreover, the pygidial glands can act like a revolving turret, allowing the beetle to shoot this acidic solution in multiple directions, even over its head! While being sprayed by this irritating compound is bad enough, the mixture of these two chemicals is exothermic, that is it produces heat, and the resulting solution can reach temperatures near 100°C. Needless to say, any would-be predator is in for a surprise. For example, when toads (genus *Bufo*) try to eat these beetles, the noxious spray causes the toad to regurgitate it and the bombardier beetle can escape unscathed: beetles have been recorded emerging unharmed 107 minutes after being swallowed. Sadly, the bombardier's explosive sprinkler isn't used for flight, but scientists are studying these beetles with the hopes of improving propulsion designs.

SOME GECKOS GET NAKED
TO AVOID BEING EATEN

TRUE OR POO? TRUE

Madagascar is an amazing place; sometimes described as "the eighth continent," this island off the coast of East Africa has incredibly high biodiversity, and it is estimated that over 90 percent of all the animals and plants that live there are found nowhere else in the world (the term for this is endemic species). Geckos of the genus *Geckolepis*, the fish-scaled geckos, are no exception—all but one described species are found only in Madagascar. These dull brown geckos seem fairly unassuming at first, other than the fact that they have pretty big scales. They have a rather unusual defence mechanism, however— when grabbed they will shed their scales, as well as the upper layer of skin, leaving the gecko 'nude' and resembling a slimy, pink, gecko-shaped sausage. This may sound painful, but their scales are specially adapted to do this, and the gecko is unharmed—bar maybe some minor embarrassment over its nudity. The scales grow back over the course of just a few weeks, leaving the gecko free to strip off again in the future should the need arise, This is not a defence strategy we would recommend for use by humans, however, as not only would it probably prove ineffective, they would also risk being arrested and charged with indecent exposure.

CUTTING AN EARTHWORM IN HALF CREATES TWO EARTHWORMS

TRUE OR POO? POO

While working in the garden it is not uncommon to see an earthworm (order *Megadrilacea*), and sometimes they can accidently meet the business end of a shovel. When this happens you might be surprised to find two wriggling earthworms where there once was one. But this doesn't mean that both ends will become a new individual, at best you'll likely still just have one. It is true that earthworms are able to regenerate tissue, but typically only the tail will grow back from the half that contains the head, which is the end nearest the thick, and usually lighter-colored, band known as the clitellum (a reproductive organ that secretes a cocoon for its eggs). The other half, unfortunately, may continue to wriggle due to latent nerves still firing, but is unlikely to grow a new head. Interestingly, in laboratory conditions earthworms that are cut asymmetrically between the clitellum and the head (no further down the body than the twenty-third segment) regeneration of the head is still possible. However, even in sterile lab conditions the part that regenerates the head may have a dysfunctional digestive system (or worse, one half of the worm may end up with two tails) making long-term survival unlikely.

TOADS GIVE YOU WARTS

TRUE OR POO? POO

Although toads and frogs may seem like very different animals, toads are just what we call some species of frog. In fact, there isn't actually any defined reasoning for which species are commonly called frogs and which are called toads, other than generally that toads are species of frog with shorter legs and drier skin (there are exceptions, see the Surinam toad, see page 19). A large number of "toad" species have bumps on their skin. These bumps are not warts, which are harmless growths caused in humans by an infection with human papilloma virus, because toads cannot catch or transmit these. We call toad's bumps "warts" because... they kind of look like warts. In reality, though, these bumps are glands that produce a toxin if the toad is distressed, which deters predators from eating them. So probably best not to go around touching (or licking, see page 91) toads if possible—unless you are helping them migrate! The common toad (*Bufo bufo*) migrates back to the same body of water it was born in to breed. Sadly, with increasing urbanization, there are often now obstacles in place, such as roads, fences, and ditches, which prevent the toads migrating to their ancestral breeding grounds. In many parts of the UK there are toad patrols, organized by local wildlife groups, which help the toads to cross any barriers on their migratory path.

WASPS ARE OUT TO GET YOU

TRUE OR POO? POO

It happens every summer, you are enjoying a nice picnic outside when an uninvited wasp (family *Vespidae*) arrives and starts buggin' you, unfortunately this can sometimes be accompanied by a painful sting. It might seem like wasps are just out to sting you and ruin your picnic, but there's a reason behind their behavior. Worker wasps spend most of their lives finding high-protein foods, like carrion or insects, to feed the colony's young. However, at the end of summer, after the last brood has been raised, the workers' job is done and they will seek out fruit and other sugary foods for themselves; if the fruit has become rotten and fermented then the wasps can become drunk and their behavior more erratic! The best way to avoid a sting is to ignore them, as swatting at them can provoke an attack and killing them can cause the release of pheromones that signals other wasps to defend the area. Scientists have found that certain fragrances containing apple or banana flavorings can also elicit this defense response. So unless you are the host species for a parasitoid wasp (see page 131), which is unlikely if you are reading this book, then wasps certainly aren't out to get you—they're just drunk and making poor life choices.

YOU CAN GET HIGH FROM LICKING A TOAD

TRUE OR POO? TRUE

All sorts of substances can get all sorts of animals high—you've probably seen pet cats get high on catnip, for example. Cats will act funny for about 10 minutes, before returning back to normal. Historically, humans have used a wide variety of plants and chemical substances to elicit this feeling—known as recreational drugs. One of the strangest mechanisms people have used to do this is undoubtedly by licking toads, in particular the cane toad (*Bufo marinus*) and Colorado River toad (*Bufo alvarius*). These toads produce the toxin 5-methoxy-N, N-dimethyltryptamine to make them unpalatable to predators. This chemical is a serotonin antagonist that binds to serotonin receptors and then releases a lot of it into the body. Serotonin is the substance that makes us feel happy, and as a result of the increase in serotonin production, people who have ingested 5-methoxy-N, N-dimethyltryptamine report a rush, as well as powerful hallucinations. Toads also produce a cocktail of other toxins, including ones that make you vomit and some that cause irregular heartbeat, seizures, and, in many cases, death. So while you can technically get high from licking a toad, you run a great risk of dying immediately afterwards. Don't do drugs, kids.

SOME NEWTS STAB THEMSELVES TO DETER PREDATORS

TRUE OR POO? TRUE

Back in 1879 a German zoologist named Franz Leydig described an unusual trait in the Iberian ribbed newt, *Pleurodeles waltl*. When startled, sharp spines appeared along the newt's sides. These aren't just any spines, though, they are actually the newt's ribs, which are stabbed through the skin as a defence mechanism. It turns out these newts physically rotate their ribs forwards, piercing their body wall in the process, in order to make them difficult for predators to swallow. If that doesn't sound unpalatable enough for a potential meal, they also secrete a slimy, toxic substance from their skin, which tastes bad and may even be potent enough to kill small animals. While it may seem counterintuitive to stab yourself from the inside out in order to avoid being eaten—as wounds mean that animals are much more likely to pick up infections—amphibians have an extraordinary ability to repair their skin. Amphibian skin also contains specialized antimicrobial peptides which provide protection against many pathogens, including bacteria and fungi. Because of the awesome regenerative powers, a number of Iberian ribbed newts have been sent into space—amazingly, the newts were found to regenerate faster in space than they do here on Earth!

FULMARS HAVE DEADLY VOMIT

TRUE OR POO? TRUE

Fulmars (two species of gull-like seabird in the genus *Fulmarus*), have a primarily seafood diet, and they are more than happy to also scavenge discarded entrails that they might find where humans are fishing. If you were to imagine the vomit produced from this diet you would likely conjure up some disgusting images, and you would be correct. Their orange and oily vomit has a putrid, fishy smell that fulmar chicks actually use as a defense mechanism. The babies will project their vomit at any intruder and the smell will understandably drive them away. Vomiting to dissuade predators is not unique to the fulmar, as chicks of other birds like the Eurasian roller (*Coracias garrulus*) will throw up on themselves, and when the parents return to find—and smell—the vomit-soaked chicks they know that danger may be near. But what makes the fulmar's vomit so special is that the oils contained within it can strip the waterproof coating from the feathers of other birds, meaning if they get wet, their waterlogged feathers can leave them exposed to harsh weather conditions or even cause them to drown.

SPECIES MISNOMERS

In the mid 1700s, Carl Linnaeus' work laid the foundation for the standardized naming system for all the living creatures that scientists still use today. Species are classified in a hierarchy, starting with three domains: *Archaea* and *Bacteria*, which can often be difficult to tell apart as both do not contain a nucleus within a membrane, but can be distinguished based on several characteristics, such as their cell membranes or metabolic pathways The third domain, *Eukarya*, contains single and multicellular organisms that have a membrane-bound nucleus. Biological organisms can be further classified based on their kingdom, phylum, class, order, family, genus, and finally species. The last two taxonomic ranks, genus and species, are combined into a unique name for each species, which is referred to as binomial nomenclature, or their scientific name. For example, the human species, *Homo sapiens*, is classified in the *Hominidae* family, which is in the order *Primates*, which are in the *Mammalia* class that is found in the phylum *Chordata*, which is within the *Animalia* kingdom that is found in the *Eukarya* domain.

This system of classification allows us to assign unambiguous names to each species; however, species also have common names, which unfortunately are not always clear in their naming conventions, nor are they always unique to a given species. Common names are based on everyday language, so they have the advantage of being easier to pronounce compared to the Latin scientific names. Because why would you say *Parastratiosphecomyia stratiosphecomyioides* when you could just say Southeast Asian soldier fly? Many common names have also been used for hundreds of years, so their use today might seem weird, and sometimes even a bit crude.

In this section you will find a collection of common names that appear to incorrectly name a given species, but is each one truly a misnomer? What exactly is a horny toad? (See page 100.) Are electric eels really electric? (See page 98.) Where can you find a slippery dick? (See page 102.) Do snot otters constantly have runny noses? (See page 104.) Just how loud is the screaming hairy armadillo? (See page 106.) Are nighthawks simply nocturnal hawks? (See page 108.) How did the cockchafer get its name? (See page 103.) Are boobies the only birds capable of nursing their young? (See page 109.) Is a giant floater something that you find in a toilet? (See page 111.) Which panda is the true panda? (See page 112.) Are mantis shrimp half mantis and half shrimp? (See page 114.) Do black widows deserve their name? (See page 116.) What exactly is coral? (See page 113.)

ELECTRIC EELS ARE A SPECIES OF EEL

TRUE OR POO? POO

Electric eels. They are eels, and they are electric, right? WRONG. Well, okay, half right, they are electric. Rather than being eels—which are fish in the order *Anguilliformes*—the electric eel is actually a type of South American knifefish (named for their knife-like body shape, not because they are used to stab things) in the order *Gymnotiformes*. All South American knifefish are capable of generating electric fields—generally of just a few millivolts—which they use to navigate and communicate in the murky waters in which they live. This helps them locate their food, usually small invertebrates that live on and in the river bed, and also locate each other. The electric eel, however, takes this electricity generation to a whole new level. They can produce electric shocks of up to 860 volts—enough to kill their prey, which is mostly invertebrates such as crustaceans, but also fish and occasionally small mammals. They can also deliver a very painful shock if disturbed by predators, or indeed humans! Shocking. This isn't the only thing their electricity is used for, though—Miguel Wattson, the resident electric eel at the Tennessee Aquarium, uses his electricity to send out regular tweets from his very own Twitter account.

THE HORNY TOAD IS A TYPE OF FROG

TRUE OR POO? POO

In another example of people naming animals totally illogically, the horny toad is not, in fact, a species of frog with horns, or even one that is particularly lustful. It is actually a type of spiny lizard. Horny toads, also known as horned lizards, are a genus of seventeen North American lizard species, *Phrynosoma*. Probably the most famous thing about these lizards is their unique method of defense—they squirt blood from a pouch below their eyes, called the ocular sinus, into the mouth of an attacking predator. Interestingly, these lizards actually use different approaches to ward off different predators. Blood squirting is reserved for canids (for example, foxes and coyotes) and felids (generally bobcats or domestic cats). If a snake or roadrunner comes along they puff themselves up to advertise their size and make themselves as difficult to swallow as possible.

So what happens if a human approaches a horned lizard? Well, not a lot really. They will probably run away, and if you catch them they may puff up and hiss a bit. The reason blood squirting is reserved for canids and felids is because horny toad blood has special compounds that bind to their taste receptors, giving the predator a very unpleasant taste in their mouth. It doesn't work on other species, though—including humans. How do we know this? Well, scientists decided to taste it. The things we do in the name of science...

YOU CAN FIND SLIPPERY DICKS IN THE OCEAN

TRUE OR POO? TRUE

Halichoeres bivittatus is a species of wrasse that is found in shallow, tropical waters of the western Atlantic Ocean—and yes, its common name is the slippery dick. They live in shallow reefs and seagrass beds close to the coast, at depths of up to 30 meters. The fish got its name for its incredible ability to avoid capture, as it easily slips out of nets and hands, and has even been reported to jump out of fish tanks. Dick was a common name, short for Richard, back when this fish was named, so presumably some poor bloke named Richard reminded someone of this fish to the extent they named it after him! Wrasse change gender throughout their lives, and the slippery dick is no exception, but as opposed to clownfish (see page 18) wrasse changes from female to male. Female wrasse are smaller, younger fish, and once they hit a certain age and size they change into males—known as protogynous hermaphroditism. Slippery dick is far from the only ridiculous common name of a wrasse species, though, other examples include splendid leopard wrasse, puddingwife, sunburnt pigfish, and carpenter's flasher. The word wrasse, in fact, comes from the Cornish word *wragh*, which means old hag. Charming.

THERE IS A "COCKCHAFER" BEETLE

TRUE OR POO? TRUE

Beetles (Order *Coleoptera*) are weird. Bombardier beetles have an explosive butt (see page 83), whereas other species pretend to be ant butts (see page 128). But even beetle names can be a little unusual. Take, for example, beetles in the genus *Melolontha*, which are commonly known as doodlebugs, May bugs, or even cockchafer. The "May bug" name is fairly straightforward, as these beetles are usually encountered in late spring or early summer, and "doodlebug" refers to these beetles' seemingly aimless flying. There is a reason for the name "cockchafer" as well, and it isn't what you are thinking—chafer simply means beetle, whereas the name cock is an old English name for an animal of large size.

Despite their humorous name, cockchafers can be agricultural pests, and not surprisingly humans don't want them around. But in 1320, the city of Avignon, in France, put all cockchafers on trial and banished them to a special reserve else they would be killed. Most likely in part due to their inability to understand the human language, the cockchafers did not comply with this decree. Some other strange beetle names are the result of a scientist with sense of humor (as scientists, we can confirm that a not insubstantial number of us do possess this characteristic). Obviously, the allure of discovering beetles within the genus *Colon* is just too great and scientists have provided such great names as *Colon rectum* and *Colon grossum*.

SNOT OTTER IS WHAT YOU CALL AN OTTER WITH A COLD

TRUE OR POO? POO

No, a snot otter is not an otter excreting an unusual amount of mucus due to contracting a winter virus. It isn't even a mammal at all, in fact, but a large aquatic amphibian. The snot otter (*Cryptobranchus alleganiensis*) has a spectacular set of names: it is also known as the hellbender, the mud devil, the ground puppy, and "old lasagne sides"—after the folded skin on its sides that someone somewhere obviously thought resembled lasagne. This giant salamander can measure up to 74 centimeters from nose to tail, and weigh up to 2.5 kg making it the third-largest salamander in the world, after the Chinese giant salamander (*Andrias davidianus*) and the Japanese giant salamander (*Andrias japonicus*). Snot otters live in shallow, swift-moving rivers in the Eastern United States— from southern New York to Georgia. They hang out under rocks during the day, coming out at night to feed on crayfish and small fish. As juveniles they have gills, but they lose these as adults and use their specialized "lasagne-like" sides (We really don't know what kind of weird grey lasagne people have been eating?!) to exchange gasses for respiration. Sadly this makes them particularly at risk of disturbance, as they need clean, well-oxygenated water to breathe and, as a result, the population has steadily been declining for decades due to habitat loss and reductions in water quality.

THE SCREAMING HAIRY ARMADILLO IS NAMED AFTER ITS PIERCING SHRIEK

TRUE OR POO? TRUE

Finally, an animal that actually has an incredibly appropriate name! The screaming hairy armadillo, *Chaetophractus vellerosus*, is a species of armadillo that lives in the Pampas region of Argentina, Bolivia, and Paraguay. They feed on insects, reptiles, small mammals, and plants, which they find in the sand by burying their heads before rotating their bodies quickly in a circle to form a hole. These armadillos are small, around 38 centimeters, with the armadillo trademark: armored plates covering their skin. These plates are formed by dermal bone, which is covered in special bony, scale-like structures called scutes. In hairy armadillos, of which there are three species, long, thick hairs sprout from between the plates and cover the animal's body. An armadillo's plates are primarily used for defence against predators, but only one species—the nine-banded armadillo (*Dasypus novemcinctus*)—can actually curl itself into a ball and protect their whole body. Most other species rely on fleeing from danger or partially burying their bodies in the ground to avoid being eaten. The screaming hairy armadillo, however, has a rather unique defence mechanism—when these armadillos are threatened they produce a loud piercing squeal that sounds a lot like a scream, and they don't stop until they are safe from danger.

NIGHTHAWKS ARE HAWKS

TRUE OR POO? POO

There are at least 238 species of hawk (family *Accipitridae*) but none of them are nighthawks, which are medium-sized birds that have short bills but long, pointed wings and long legs with a plumage coloration that resembles bark or leaf colors. Meanwhile, the bird called the nighthawk isn't even completely nocturnal, they can also be active in the early morning or late evening, feeding on flying insects. Nighthawks are a part of the family *Caprimulgidae*, which are collectively known as the nightjars (spoiler alert: these are neither jars nor only found at night, the name comes from the jarring sound that the male makes while the female is brooding). Nightjars have their own misnomer, these birds are also called "goatsuckers," in fact the name *Caprimulgidae* means "milker of goats." The incorrect belief that nightjars would come at night to suck the milk from goats, just like the porcupine quill myth (see page 76), can be found in the writings of Aristotle, but unlike the porcupine, he knew it to be false. However, this myth has persisted and can be found in text after Aristotle, even as recently as the 1800s! But nightjars are uninterested in milk—goat or otherwise—as they are insectivores. Likely, humans saw nightjars flying near their livestock, because that is of course where the best-flying insects are, and so the myth of the goat-sucking bird was born.

BOOBY IS A BIRD

TRUE OR POO? TRUE

The word "booby" when referencing a bird can be applied to six of the ten species of seabird in the genus *Sula*. Possibly the best known is the blue-footed booby (*Sula nebouxii*) as their bright blue feet make them readily identifiable (at least part of the name is clearly descriptive!). We can assure you that the name booby does not mean that these birds are able to nurse their young, as in mammals. The name booby is in fact descriptive, it just isn't obvious as it is a reference to their behavior. The word booby is derived from the Spanish word *bobo*, which means silly or not smart. One explanation is that sailors would call them boobies because of these birds' propensity for landing on their ships and being easily captured for food. An alternative explanation for the name is based on the boobies' awkward and wobbly walk. While the boobies' nesting behavior doesn't have anything to do with their name, in a book filled with poos, we would deeply regret not mentioning that the blue-footed booby also uses poo to help build its nest, and to keep cool by defecating on its legs and feet. What a booby.

GIANT FLOATER IS A TYPE OF MUSSEL

TRUE OR POO? TRUE

If you thought that we would write a book with the word "poo" in the title and not include the giant floater mussel (*Pyganodon grandis*), you would most assuredly be wrong. It is possible that the giant floater's color, tan to dark brown, superficially resembles feces, but that's not exactly how it got its common name. The giant floater is a large mussel, growing up to 25 centimeters. Like other mussels, this species is a filter feeder and can be found at the bottom of various bodies of water such as streams or lakes. That it typically does not float doesn't mean that it can't, though. The giant floater has a relatively thin shell—they are sometimes called papershells, and trapped air bubbles can cause them to float to the surface. For reasons that are uncertain, the giant floater has also been called hogshell or slopbucker.

But having multiple names can be problematic—how do we know what we are studying if we don't use the same name? This is especially true for mussels, due to the high variation in their morphology, or appearance. For example, the swan mussel (*Anodonta cygnea*) has a wide geographic range and can be found in a variety of habitats. As such this species has been called by at least 549 different names ("swan mussel" is definitely better than "giant floater," though).

THE RED PANDA IS A PANDA

TRUE OR POO? POO

If you've ever seen a red panda (*Ailurus fulgens*) you would probably guess correctly that the "poo" for the red panda's name is in reference to "panda" and not the word "red." Red pandas are not pandas, which are bears, but rather are more closely related to skunks, raccoons, and weasels. While red pandas and giant pandas (*Ailuropoda melanoleuca*) are only distantly related within the same order, *Carnivora*, they do share one characteristic: their diet is primarily composed of bamboo. Interestingly, the word "panda" appears to have first been used in reference to the red panda in the mid-1800s, and was possibly derived from the Nepali phrase for "bamboo footed" or "bamboo claw." It was only later applied to the giant panda given their superficial resemblance in the early 1900s, which is now formally the only extant, or living, true panda (though in a way it's the not-red panda that's not a panda. Confused?). Unfortunately, the giant panda has now completely co-opted the panda name and alternative names for the red panda are still misnomers, such as red bear-cat or cat-bear, and fire fox, as it is not a bear, cat, nor fox and it most definitely is not on fire.

CORAL IS A TYPE OF ROCK

TRUE OR POO? POO

If you didn't know that coral are marine animals, don't worry, you aren't alone—philosophers, naturalists, and scientists for over 2,000 years have incorrectly classified them as plants or even minerals. In fact, the name coral has roots in either the Hebrew or Arabic words for small pebble or stone! Corals are actually in the class *Anthozoa* and are made up of a colony of small individual organisms known as polyps that are essentially stomachs with a mouth at the end, and which are lined with tentacles. Coral can feed on dissolved molecules, zooplankton, or even small fish, using their stinging tentacles to immobilize their prey. Although coral can be soft, most people are familiar with hard coral—those that make up the diverse coral reef ecosystems like the Great Barrier Reef. Hard corals have a skeleton that is created through deposits of calcium carbonate and are home to photosynthetic dinoflagellates known as zooxanthellae, which supply the coral with oxygen and nutrients as well as give the reef their vibrant colors. Unfortunately, warming temperatures threaten these species-rich ecosystems, as high temperatures cause the coral to bleach. Where the zooxanthelle are ejected, and if temperatures do not fall quickly enough, the coral dies. Scientists are trying to find ways to seed reefs with captive-bred corals, but cutting down on carbon emissions is certainly the key to their survival.

MANTIS SHRIMP ARE SHRIMP

TRUE OR POO? POO

Mantis shrimp are many things; they are at least 450 different species, they can be visually stunning, they are fierce predators and evolutionary marvels, but they are not shrimp—and just to clear things up, they aren't mantises either. Shrimp are in the order *Decapoda*, which also includes crayfish, lobsters, crabs, and prawns, whereas mantis shrimp are classified in the order *Stomatopoda*. These two orders can both be found in the class *Malacostraca*, which also includes isopods. Mantis shrimp have a second pair of appendages on their thorax—the middle portion of their body—which come in two varieties: spears and smashers. Spearing mantis shrimp (like the zebra mantis shrimp, *Lysiosquillina maculata*) ambush their prey and quickly impale them with their sharp barbed appendages and bring them back into their burrows. Smasher mantis shrimp (like the peacock mantis shrimp, *Odontodactylus scyllarus*), have club-shaped appendages that they use to rapidly "punch"... pretty much everything at a speed of over 50 mph and a force of up to 155 kilos! Smasher mantis shrimp can be a bit of a nuisance for humans; they can break their glass aquaria or may simply kill their tank cohabitants. Smashers are also known as "thumb-splitters"—for reasons that should be painfully obvious.

BLACK WIDOW FEMALES EAT THEIR MATES

TRUE OR POO? POO (SORT OF)

Animals often have an undeserved bad reputation, and the black widow is probably one of them. Despite their name being based on the idea that the female black widows devour the male once they have mated, it turns out that this well-known "fact" is mostly a myth—or at least an exaggeration. There are thirty-one different species of widow spiders (genus *Latrodectus*), which include black widows, brown widows, and red widows. For two species, the brown widow (*Latrodectus geometricus*) and Australian redback (*Latrodectus hasselti*), eating their partner is obligatory; in fact, males will literally offer their abdomen up to the females to snack on while they mate. This might seem weird, but it actually gives him the best chance of fathering as many offspring as possible—males who are eaten during mating are able to copulate for longer, as the female is distracted by her meal, and as a result they sire more offspring than those that are not consumed. In most widow species, including American black widows, however, males actively try to approach females that have already eaten to mate with, and therefore the male is usually left to go about his business after copulation. Widows are certainly not the only animals with a habit of snacking on their partner during sex. Male praying mantises, for example, have a one in four chance of being devoured by their mate during mating, and male mantises make up over 60 percent of the females' diet in the mating season.

WEIRD PLACES TO CALL HOME

A question that many ecologists seek to answer is, where do animals live? The position within an environment that an animal occupies is known as its niche, and this is linked to its diet, its predators, and any changes that it may make to the ecosystem in which it lives. On a smaller scale, the environment that an animal inhabits is known as its habitat—for example, the savannah, tundra, forest, or a wildflower meadow. These are places where an animal can find food (see page 24), shelter, and other individuals of the same species with which to socialize and mate (see page 1).

Animals live in an amazing array of habitats. These can be from the tiniest scale, such as mites that live on beetles, to enormous scales—for example, the humpback whale, which migrates over 10,000 miles across the world's oceans each year. Go almost anywhere on Earth and you will find thriving animal communities, whether that is in the cities of London, New York, or Beijing; the vent of a deep-sea volcano; the ice floes of Antarctica; or areas of the Amazon untouched by anyone but the native peoples that live there. Within those communities animals do some amazing things—they break down decomposing matter, help keep waterways clean, regulate and pollinate plant communities, or even help other animals to breed. Despite the fact that animals live in an incredible variety of places, many myths still persist around where they do in fact live.

The truth, however, is often even more remarkable than reality. The pearlfish hides out in sea cucumber cloacas to stay safe from predators. Tracheal mites live in a bee's respiratory system—they may seem tiny to us but they are actually a sixty-eighth of the size of the bee (human mites are around one-4,125th of the size of a human). Imagine having something that big crawling through your airways! There are even specialized bat flies that live only on bats (not a misnomer, see page 96), many of which only live on certain bat species. These may not seem like fun places to live for you and me, but to these species it is home sweet home.

In this section you will learn about some traits that seem incompatible with a species' habitat (sharks, see page 120), some astonishing animal relationships (frog–tarantula, see page 124; ant-butt, see page 128) and some rather less-friendly relationships (fish–isopod, see page 126; wasps, see page 131; barnacle-crab, see page 134). You'll also discover some strange and surprising places where animals do and don't live (megalodon, see page 130; rodent-pests, see page 123) including on and in the human body (earwigs, see page 136; face-mites, see page 137), and just how many rats you are sitting near as you read this book (see page 122).

SHARKS LIVE IN THE SEA BUT DROWN IF THEY STOP SWIMMING

TRUE OR POO? POO

Telling yourself to keep swimming might be something you say when you want to persist through hardships, but is this something that sharks need to say to themselves in order to stay alive, as many people think? In order to extract oxygen from the water, sharks must pass water over their gills, but they don't all necessarily need to swim to accomplish this. Some sharks, such as nurse sharks (*Ginglymostoma cirratum*), can "breathe" without swimming using buccal pumping, which uses buccal muscles (kinda like their cheeks) to pump water into their mouths and over their gills. However, other species, like the great white (*Carcharodon carcharias*) do not have this ability, and pass water over their gills by swimming quickly with their mouths open, which is known as ram ventilation. But that doesn't mean that ram ventilator sharks never sleep—even sharks need to rest! For example, researchers have found sharks motionless inside an underwater cave (now known as the cave of sleeping sharks)—likely the high dissolved oxygen in these waters allowed their rest. Moreover, scientists have found that swimming is coordinated by a shark's spinal cord rather than their brain, so it is possible for them to unconsciously swim.

YOU ARE ALWAYS WITHIN 6 FEET OF A RAT

TRUE OR POO? POO

Unless you keep a rat in your home as a pet, or are reading this in the Karni Mata Temple, in India, which is home to around 25,000 rats that are revered as reincarnations of Karni Mata children, you probably aren't that close to a rat right now. Rats like humans because we have food. Not surprisingly, we have a history with these rodents. Rats can transmit diseases to humans, the most famous of which is likely the bubonic plague of the 1300s known as the Black Death. However, scientists have recently found that rats may have been framed and that human fleas and lice were a more likely cause of the spread of the disease-causing bacterium.

The myth that rats are always close to you, though, may have been started in the UK in the early 1900s when W. R. Boelter, who surveyed people in the English countryside for his book *The Rat Problem*, used those surveys to estimate that there were as many rats as humans at that time—around 40 million. Modern estimates put the rat population in this area closer to 10 million, and while that may seem high, 70 percent of these rats are found in rural areas and not in the cities, as some might believe. In some places you may have to look far and wide to even find a rat—Alberta, Canada is considered rat-free, at least of brown rats (*Rattus norvegicus*), due to an aggressive rat-control programme that has been in force since the 1950s!

RODENTS ARE INTRODUCED PESTS AND ONLY FOUND IN UNCLEAN PLACES

TRUE OR POO? MOSTLY POO

A pest can be any living organism that attacks crops, livestock, or other food sources, or is just generally considered a nuisance to humans. Clearly some rodent species can fit this definition—by spoiling our food (you don't want to know how much rodent hair is in your food), by transmitting diseases to us or livestock, or simply being unsightly in and around human dwellings. Unfortunately, it is true that humans have moved some rodent species, like the black rat (*Rattus rattus*, see page 122), into new areas and their presence in these new habitats, especially islands, have caused the extinction of many species of native reptiles, birds, or mammals. But rodents get a bad rap. Nearly 40 percent of all mammal species are rodents. Native rodent species can be found on all continents except Antarctica and in diverse habitats from deserts to the tundra, underground as well as in trees. Rodents can perform a variety of essential ecological roles, such as plant pollinators or seed dispersers, provide an important link in food webs, or (like the beaver, see page 69) can act as habitat engineers. Perhaps the greatest ecosystem service belongs to the largest living rodent, the capybara (*Hydrochoerus hydrochaeris*), likely due to this rodent's calm demeanor they can often be found with another animal (often birds) sitting on top of them, providing a nice resting spot or a tasty insect meal.

FROGS LIVE WITH SPIDERS

TRUE OR POO? TRUE

Mutualisms occur in nature when individuals of two different species interact and both species benefit from this interaction—an obvious example of a mutualism are plants and their pollinators (like the honey bee, see page 59) wherein the plant gains reproductive benefits through cross-pollination and the pollinator receives food. A rather unlikely mutualistic partnership is mygalomorph spiders (infraorder *Mygalomorphae*) and microhylid frogs (family *Microhylidae*). Microhylid frogs are small (hence the name 'micro'), up to ten times smaller than the spiders with which they cohabit, making them unlikely roommates! Not only do mygalomorph spiders live alongside another species that could just as easily be lunch, but they may also recognize their amphibious pets using chemical cues. Spiders have even been observed grabbing and examining frogs with their mouthparts but not eating them. Keeping frog pets is helpful to these spiders as the frogs will eat ants, which would otherwise prey on their eggs, while the tiny frog receives protection from predators, as well as few easy meals of invertebrates that are attracted to the spider's eggs or the remains of the spider's meals. Though they seem like an unlikely pairing, frog and spider mutualisms have been observed in Peru, Sri Lanka, India, and Mexico!

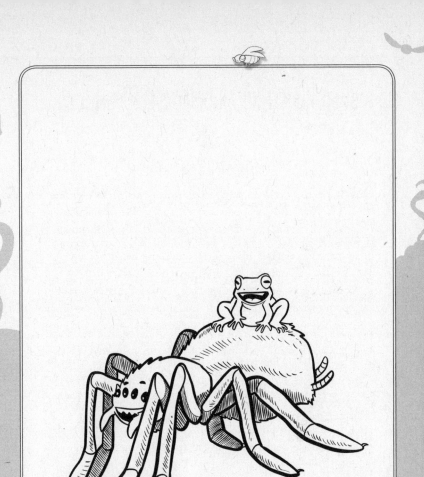

SOME FISH HAVE PARASITES FOR TONGUES

TRUE OR POO? TRUE

Having a giant woodlouse instead of a tongue may sound like something from a nightmare (and we apologize if this is now in your nightmares), but that is the reality for many fish. There are many parasitic isopods (think ocean-dwelling woodlice, basically) that inhabit fish mouths, but one species in particular takes it to an especially gruesome level. *Cymothoa exigua*, or the tongue-eating louse, is a species of isopod which basically does what it says on the tin: eats fish tongues. In particular the tongue of the rose snapper, *Lutjanus guttatus*. The female of the species enters in through the fish's gills, makes her way into its mouth and grasps onto the tongue with her seven pairs of hooked legs. She then inserts her special sucking mouthparts into the base of the fish's tongue and slowly sucks it dry, causing the tongue to eventually drop off. Next, she proceeds to attach herself in its place. These lice fully replace the fish's tongue, being used to grind up food against the roof of the mouth. The female even uses the fish's mouth as a place to raise her young, releasing juvenile isopods out into the environment once her eggs hatch. Talk about leaving a bad taste in your mouth...

THERE'S A BEETLE THAT PRETENDS TO BE AN ANT BUTT

TRUE OR POO? TRUE

Blending in with your surroundings is an excellent tactic to avoid predators. Some insects, like the aptly named giant leaf insect (*Phyllium giganteum*) have evolved a leafy appearance, whereas others, like the peppered moth (*Biston betularia*) have colors that allow them to go unnoticed while on tree bark. But one beetle, *Nymphister kronaueri*, is able to blend in with another animal—well, to be more specific, another animal's butt. These beetles use their mandibles to attach themselves to individual army ants (*Eciton mexicanum*) between the petiole and postpetiole—the narrow part of the ant's body between their thorax and abdomen. The size, shape, color, and even texture of the beetle's body is very similar to the ant's abdomen, which allows it to go undetected when viewed from above. Hitchhiking on another animal is known as phoresy, and while using an ant butt as your mode of transportation may seem rather unsavory, for this beetle using your own energy to get around, all the while running the risk of being eaten, truly stinks!

MEGALODON IS STILL ALIVE, HIDING OUT SOMEWHERE IN THE DEEP OCEAN

TRUE OR POO? POO

Megalodon was an enormous prehistoric species of shark that is known from its fossilized teeth and vertebrae. It is thought to have been between 10 and 18 meters in length when fully grown, over twice the length of a large great white shark. Megalodon teeth are over 11.5 centimeters wide—great for tearing into the small marine mammals on which it is believed to have fed. The ocean is a huge place, and it is estimated that 95 percent of the ocean floor remains unexplored. So there is no reason why a 15-meter-long prehistoric shark couldn't still be hiding out there somewhere, right? People have speculated that this might be the case, but while it might be cool to think that there is a gigantic species of shark hanging around undetected in our oceans, we can definitively say that, due to the lack of sightings and megalodon remains, this giant shark is no longer with us and hasn't been for well over a million years. If there is one thing we know about large marine animals, it is that when they die at least parts of them wash up on the shore—which is why we find a lot of dead whales around the coasts. On top of this, megalodon was a coastal species—it lived in a part of the ocean that is very well explored. This isn't to say there aren't mysterious deep sea creatures out there. Very little is known about beaked whale species, one species of which—the Cuvier's beaked whale—has been recorded diving to depths of over 3000 meters.

SOME WASPS START LIFE LIVING INSIDE OTHER INSECTS

TRUE OR POO? TRUE

It sounds like something from a horror film: wasps that lay their eggs inside other animals, which then hatch and devour them from the inside out, but it's a common occurrence in nature. There are over 600,000 species of parasitoid wasps parasitizing hosts with their offspring around the globe. The female wasp has a long, sharp ovipositor—the tube through which she lays her eggs. She will seek out her host species, which is specific to each species of wasp, and stab it, injecting her eggs into her host's body cavity. The wasp larvae then hatch out and eat their host from the inside. These larvae, when inside the host, can even change the animal's behavior to benefit them. The host of the Costa Rican parasitic wasp *Hymenoepimecis argyraphaga* is *Leucauge argyra*, a species of orb-weaving spider. When a spider is parasitized by this wasp, and the larvae are ready to emerge from its body, the spider spins a special reinforced web. The wasp larvae build their cocoon on this web once they eat their way out of their host. Luckily for us, parasitoid wasps only lay their eggs in other insects. In fact, many of these wasps are incredibly beneficial to humans—they lay their eggs in crop pests, keeping their populations in check and preventing them devouring our food.

SOME FLIES SPEND THEIR LIVES ON POO

TRUE OR POO? TRUE

The yellow dung fly, *Scathophaga stercoraria*, as its name might suggest, is a striking yellow species of fly that lives throughout much of the Northern Hemisphere... on poo. To you and me, a pile of poo might not seem like the ideal place to find a meal or raise our young, but to a male yellow dung fly, it is perfect. Males hang out on the dung—eating smaller fly species, waiting for the females. When a female comes in to lay eggs, males fight fiercely to mate with her. After mating, the eggs are deposited in the poo and she flies away, with the offspring left to mature inside the sloppy feces. In fact, cow poo is host to a whole variety of insect life, and each species has a different way of getting their offspring into that dung; some lay eggs, others deposit live young—some specialize in laying in wet pats and others prefer them when they are hard and dried. Some species of fly are so speedy at getting their eggs into the steaming poo that they lay their eggs in it before it even hits the ground. The female horn fly, *Haematobia irritans*, lives on cows and sucks their blood. Once the cow starts to defecate she quickly dashes in to deposit her offspring into the feces literally as it leaves the cow's butt!

133

SOME BARNACLES CAN TAKE OVER THE MINDS, AND BODIES, OF CRABS

TRUE OR POO? TRUE

When you think of barnacles, you probably think of the small white-shelled animal found around our coasts, clinging to rocks and the like. These tiny filter-feeding crustaceans are relatively harmless, other than causing a bit of a nuisance by attaching themselves to the bottoms of boats. There is one type of barnacle, however, that can subject you to terrifying mind control! If you are a crab, that is. *Sacculina carcini*, also known as the crab hacker barnacle, float around in the ocean as larvae. Once they find a suitable crab host, they attach to it and begin to mature. They push part of their body inside the crab, and this begins to grow, putting out tendrils throughout the crab's tissues, absorbing nutrients and allowing the barnacle to control the crab's behavior. Generally crabs keep their eggs under a specialized tail, keeping them oxygenated and safe from predators until they are ready to hatch. Instead of raising their own young, crabs infected with *Sacculina carcini* end up raising the barnacles' offspring, which it produces in a pouch under the tail. Even male crabs, which usually don't raise any eggs, are forced to do this—the barnacle cuts off the blood to their gonads, making them display more female traits that are optimal for raising the maximum number of mind-controlling barnacle babies! It seems mind control is real, after all, at least if you are a crab.

EARWIGS LAY EGGS IN YOUR EARS

TRUE OR POO? POO

Earwigs, insects of the order *Dermaptera*, have been on this planet for at least 208 million years. Sadly, since Roman times they have been given a bit of a bad reputation. Maybe people believe that earwigs actively seek out people's ears to burrow into and lay their eggs. This is most definitely a myth, and it is thought that this story was started by Pliny the Elder, who also wrote a lot of other things that we today know to be wrong. For example, he claimed that women were never left-handed, that coral isn't an animal (see see page 113), and that you could cure toothache by injecting earwigs boiled in oil into your ear (please don't try this). Earwigs do like moist, dark places, which may make an ear seem like an appealing place for them, but they have a strong preference for natural habitats such as in soil and under tree bark, and an earwig has never been reported to have laid its eggs inside a person's ear canal. Although earwigs have, on very rare occasions, been found in people's ears, this hasn't happened any more often than with other insects, such as moths or spiders. One piece of advice from the Romans for dealing with an insect in your ear was to spit in it, but we would recommend instead seeking advice from a medical professional, who can easily remove it with a pair of tweezers.

THERE ARE MITES ON YOUR FACE

TRUE OR POO? TRUE

You have mites on your face. Right now. You probably have mites all over your body. But before you run to the bathroom to scrub your face you should know that these mites are everywhere, and will be back as soon as you come in contact with another human. *Demodex folliculorum* and *D. brevis* are in the same class as scorpions, ticks, spiders, and harvestmen—*Arachnida*. *D. folliculorum* lives in and around larger pores and hair follicles, whereas *D. brevis* can be found in sebaceous glands (which are responsible for making your hair oily). Because these structures are found more readily on your face, so are the mites. It might help to know that our face mites are not parasites but rather are considered, at worse, commensals, which means they obtain nutrients from us but do us no harm. Some research, however, suggests that they "mite" be beneficial— like many of the microscopic organisms that call your body home (see see page 53)—and may clean our skin of dead cells or harmful bacteria. These mites don't have anuses or a similar structure from which they excrete solid waste, which means they aren't constantly pooing on you! However (and some of you may want to stop reading now), that means when they die—their lifespan is a maximum of 16 days—they are totally filled with poo, which is all released in one go upon their demise. Have a nice day!

GLOSSARY

Abdomen – In mammals, the part of the body that mostly contains digestive organs such as the intestines, while in arthropods this is the last, or posterior, section of their body after the thorax.

Adaptation – An alteration to an organism that suits its environment.

Altruism – An animal's behavior that negatively impacts itself while having a positive impact, or benefit, on another.

Ancestral – Referring to a place where an animal originated.

Anthrax – A disease caused by a bacterium, *Bacillus anthracis*, that affects the skin and lungs.

Antimicrobial – Any substance that kills microorganisms.

Appendage – Any projection from a living organism that has a specific function.

Archaea – One of the three domains of living organisms that are morphologically similar to bacteria but also share genetic similarities with eukaryotes.

Avascular – Tissue that does not have blood vessels.

Autotomy – Discarding of a body part, often used as a defence mechanism.

Botulism – A type of food poisoning that can occur in preserved foods or canned meats that are not stored or sterilized properly—caused by three bacteria in the *Clostridium* genus.

Brooding – Sustaining a proper environment, like temperature or humidity, for eggs by sitting on them or maintaining a close relationship with them.

Buccal – Refers to the cheeks or generally the mouth.

Carrion – Flesh from dead and decaying animals.

Carboniferous – From 359.2 to 299 million years ago during the late Paleozoic Era. Marked by the formation of coal deposits and the evolution of eggs in which the embryo is surrounded by a series of fluid-filled membranes.

Carnivorous – Animals that feed on the flesh of other animals.

Clitellum – The reproductive segment of worms. Appears as a raised band or collar.

Coprophagy – Eating of feces.

Copulation – Joining together during sexual reproduction.

Dermal Bone – Bone that is formed within the dermis of the skin.

Diaphragm – The main muscle of breathing, in most mammals it separates the thorax and abdomen.

Dinoflagellate – A unicellular protist found in both marine and freshwater habitats that contain two flagella or threadlike structure that enables the organism to swim.

Embryo – An organism that is in the earliest stages of development.

Entrails – Intestines and other internal organs that have been removed.

Esophagus – A long muscular tube that connects the mouth to the stomach.

Eukaryote – One of the three domains of life marked by cells that contain a membrane-bound nucleus.

Exoskeleton – In invertebrate animals, the external covering that provides support and protection.

Exothermic – A chemical reaction that releases heat.

Extant – Species that are still living.

Feces – Solid waste formed by the remains of undigested food.

Fertilize – The union of an egg with sperm.

Genome – The genetic material (DNA) of an organism.

Gestation – The process of an individual's development that occurs after fertilization until an organism is born.

Gonads – Organs that produce the reproductive cells or gametes.

Herbivorous – Animals that feed on plant material.

Hermaphrodite – An organism that possesses the gonads of both sexes.

Hectocotylus – A modified appendage of male octopuses that contains sperm.

Hind Gut – The end part of the gut or intestines.

Hydrogen Peroxide – A chemical compound often used as a bleaching agent. Mixes with hydroquinone to form the explosive defence of the bombardier beetle.

Hydroquinone – An aromatic phenol—that is, a smelly organic compound—produced in bombardier beetles.

Insectivore – An animal that eats only insects.

Invertebrate – An animal without a backbone.

Iridophore – A type of reflective-pigment-containing cell found in chameleons.

Mandible – Appendages near an insect's mouth. Also the lower jaw of vertebrates.

Midgut – The middle part of the intestines or the middle of an insect's digestive system.

Millivolt – One-thousandth of a volt.

Mite – A small arachnid with four pairs of legs.

Mate – An animal's partner. Also the process of copulation.

Monogamy – Possessing only one mate at a time.

Morphology – A specific shape or form.

Nasal Cavity – The large air-filled space behind the nose or beak.

Offspring – The young, or progeny, of an organism.

Oviduct – The tube which eggs pass down after leaving the ovary— the part of the body that produces eggs in most animals

Papillae – A small protrusion form part of an animal's body.

Parasite – An organism that lives on or in another organism and gains benefits at the expense of its host.

Parasitize – To infect or live in another organism as a parasite.

Pathogen – An organism that causes disease.

Pepsin – An enzyme that breaks down proteins.

Petiole – A slender stalk between the thorax and abdomen of an ant or wasp. Also the stalk that joins the leaf to the stem in a plant.

Pharangeal – In the pharynx.

Pheromone – A chemical released into the environment that affects the behavior of other organisms.

Phoresy – A type of symbiosis where one animal is transported by its host.

Pliny the Elder – A Roman author, naturalist, philosopher and early commander of the Roman Empire who was very wrong about a lot of things.

Pompeii – An ancient city near Naples in Italy. It was buried in the eruption of Mount Vesuvius in the year AD79.

Postpetiole – The second segment of the pedicel (the part that connects the abdomen to the rest of the body) of some ants.

Prostoglandin – A type of lipid (a molecule that does not easily dissolve in water) that prevents the production of stomach acid in gastric breeding frogs.

Protists – Often defined as animals in the kingdom Protista. A eukaryote that is not an animal, plant, or fungus. They are hard to describe as they don't have a whole lot in common. Grouping living organisms is hard.

Protogynous Hermaphroditism – Organisms that are born male and turn female at some point in their life.

Pus – The yellow or white liquid produced when the body has an infection, which is found in spots.

Pygidial Gland – A specialist gland fund in some beetles that produces chemicals.

Sediment Core – A cylindrical sample of soil, sand or other sediment.

Sequential Hermaphroditism – When a species changes sex at some point in its life.

Serotonin Antagonist – A drug used to inhibit the action at serotonin.

Sex – A category an organism falls into based on its reproductive function. For most species this is male and female, but some species only have one sex or many sexes.

Symbiosis – A close long-term interaction between two organisms.

Tetanus – A disease also known as lockjaw that causes muscle spasms.

Thorax – Either the part of the body of a mammal between the neck and the abdomen or the middle section of an insect's body.

Toxicity – How toxic or poisonous something is.

Trachea – Windpipe.

Unpalatable – Tastes bad.

Urbanization – The process by which an area becomes more urban (like a city) and less rural.

Volt – A unit of measurement of electrical potential.

Zooplankton – Microscopic animals found in our oceans and waterways.

Zooxanthellae – Specialized single-celled dinoflagellates (a type of planktonic protist) that live in symbiosis with animals such as corals and jellyfish.

ABOUT THE AUTHORS

Dani Rabaiotti is a PhD researcher and science communicator at the Zoological Society of London and University College London. She is currently researching the impact of climate change on African wild dogs, and spans various fields, including zoology, quantitative ecology, and macro ecology. While her research has involved fieldwork in Kenya, in reality Dani actually spends most of her time in London at a computer, coding—which may be somewhat less glamorous, but is just as fun and considerably less stinky. She considers it of paramount importance that everyone is aware of the fact that African wild dogs have the best ears in the animal kingdom. Dani will be finishing her PhD in 2019.

Nick Caruso is currently a postdoctoral associate in the Department of Fish and Wildlife Conservation at Virginia Tech. His current research focuses on ecology and conservation of Appalachian salamanders and herpetofauna in the Florida panhandle. When he is not researching amphibians and reptiles or writing books about farts and poos, he's mountain biking with his wife, playing racquetball or soccer, or cuddling his cats.

ABOUT THE ILLUSTRATOR

Ethan Kocak is an artist and illustrator best known for the online comic series *Black Mudpuppy* and various science-related art projects including the *New York Times* bestseller *Does It Fart?* and TV presenter and biologist Ben Garrod's dinosaur series *So You Think You Know About Dinosaurs?* He lives in Syracuse, New York, with his wife, son, and collection of rare salamanders.

CONTRIBUTORS

Thank you to everyone that pointed us in the direction of the Trues and Poos included in this book, and to the experts that fact-checked our entries! You can find them at the Twitter handles below:

@_glitterworm
@alexevans91
@AlexSlavenko
@AnnaDeyle
@annfro
@ashaelr
@AVScards
@barreleyezoo
@battragus
@becomingcliche
@berlinbuggirl
@birdnirdfoley
@CanopyRobin
@Cataranea
@claireasher
@ConnectedWaters
@DiamondKMG
@drmichellelarue

@dllavaneras
@EntoLudwick
@Fuller_Si
@hammerheadbat
@JessieAlternate
@Julie_B92
@MarkScherz
@MazHem_
@mbystoma
@mdahirel
@NadWGab
@NatickBobCat
@rickubis
@TeenyannB
@temptoetiam
@TheLabandField
@WhySharksMatter